Ones and Zeros

Understanding Boolean Algebra, Digital Circuits, and the Logic of Sets

John Gregg

IEEE Press Understanding Science & Technology Series
Dr. Mohamed E. El-Hawary, *Series Editor*

The Institute of Electrical and Electronics Engineers, Inc., New York

IEEE Press
445 Hoes Lane, P.O. Box 1331
Piscataway, NJ 08855-1331

Cover design: Caryl Silvers, *Silvers Design*

Technical Reviewers

Sherif Embabi, *Texas A&M University*
Stam Kartalopoulos, *Lucent Technologies, Bell Labs Innovations*
Richard S. Muller, *University of California, Berkeley*
Kevin D. Taylor, *Purdue University at Kokomo*
Neal S. Widmer, *Purdue University, EET Department*

ONES AND ZEROS

IEEE Press Understanding Science & Technology Series

The IEEE Press Understanding Series treats important topics in science and technology in a simple and easy-to-understand manner. Designed expressly for the nonspecialist engineer, scientist, or technician, as well as the technologically curious—each volume stresses practical information over mathematical theorems and complicated derivations.

Books in the Series

Blyler, J. and Ray, G., *What's Size Got to Do With It? Understanding Computer Rightsizing*

Deutsch, S., *Understanding the Nervous System: An Engineering Perspective*

Evans, B., *Understanding Digital TV: The Route to HDTV*

Gregg, J., *Ones and Zeros: Understanding Boolean Algebra, Digital Circuits, and the Logic of Sets*

Hecht, J., *Understanding Lasers: An Entry-Level Guide,* 2nd Edition

Kamm, L., *Understanding Electro-Mechanical Engineering: An Introduction to Mechatronics*

Kartalopoulos, S. V., *Understanding Neural Networks and Fuzzy Logic: Basic Concepts and Applications*

Lebow, I., *Understanding Digital Transmission and Recording*

Nellist, J. G., *Understanding Telecommunications and Lightwave Systems,* 2nd Edition

Sigfried, S., *Understanding Object-Oriented Software Engineering*

To Susan, without whom this book would have been finished a lot sooner.

This book may be purchased at a discount from the publisher
when ordered in bulk quantities. Contact:

IEEE Press Marketing
Attn: Special Sales
445 Hoes Lane, P.O. Box 1331
Piscataway, NJ 08855-1331
Fax: 1-732-981-9334

For more information on the IEEE Press,
visit the IEEE home page: http://www.ieee.org/

Printed in the United States of America

10 9 8 7 6 5 4 3 2 1

ISBN 0-7803-3426-4
IEEE Order Number: PP5388

Library of Congress Cataloging-in-Publication Data

Gregg, John.
Ones and zeroes : understanding Boolean algebra, digital circuits, and the logic of sets / John Gregg.
p. cm.
Includes bibliographical references and index.
ISBN 0-7803-3426-4
1. Electronic digital computers—Circuits—Design. 2. Logic, Symbolic and mathematical. 3. Algebra, Boolean. 4. Set theory. I. Title.
TK7888.4.G74 1998
511.3′24—dc21 97-34932
CIP

Contents

Before We Begin

This book is primarily about Boolean algebra, a remarkable system of mathematical logic through which its inventor, George Boole, sought to characterize all of human intelligence in precise symbolic form. He failed, as have all who came after him. He did succeed, however, in accomplishing two things. First, he created the entire field of symbolic logic, and thereby rescued logic as an active field of inquiry from a 2000-year lull that began when Aristotle died. Secondly, many decades after his death, Boole's system of symbolic logic was used as the conceptual basis for certain types of electrical relays and switches. We now know these types of devices by the name ''digital circuits,'' and they form the basis of all modern computing machinery.

It is important to understand that a working computer consists of both hardware and software, and the digital circuits discussed in this book comprise only the former. Software consists of commands, written by human programmers, that reside in a computer's memory. It is the job of the hardware to fetch those commands in the correct sequence and act on them. Writing software is by no means a trivial exercise, and there are many good books available on the subject of software engineering, although this book is not one of them.

Boole's system and the circuits built on it that are described in these pages are surprisingly simple. Furthermore, this book is about logic (often as applied to electrical engineering), but not about electrical engineering per se. You will not find words like watt, ohm, capacitance, or resistance in this book. I assume no more than a high-school level of familiarity with general mathematical concepts.

I encourage you to read through the exercises as you read the book. There are some interesting wrinkles in them that extend the ideas presented in the text. Remember, of course, that the answers are all in the back of the book.

If this book strikes you as at all interesting, I would also encourage you to glance through the reading list in the back. It is categorized by topic and, among other things, lists many introductory electrical engineering and digital circuits texts. These texts tend to cover much of the same material that this book does, but in more detail and depth. The reading list also contains several readable and fascinating books about mathematics in general and the history of its development.

John Gregg

0

Number Systems and Counting

The focus of this book is the branch of mathematical logic known as Boolean algebra or Boolean logic. Although we will spend considerable time exploring Boolean algebra in its various forms, first we should spend time thinking about numbers and the ways in which we represent them by using symbols.

0.1 NUMBERS: SOME BACKGROUND

A **numeral** is a single symbol that represents a quantity or number. A **number system** is a way of assigning combinations of numerals to different quantities. The act of assigning combinations of numerals to different quantities through the use of a number system is called **counting.** Human beings happen to have 10 fingers, five on each hand, which has resulted in the use of the **base 10** or **decimal** number system by most cultures. In fact, the word **digit** comes from the Latin *digitus,* meaning finger.

The decimal system has 10 numerals: 0, 1, 2, 3, 4, 5, 6, 7, 8, and 9. We can express any number we want by using different combinations of these 10 numerals within the base 10 number system. Some cultures, notably the Mayas, Celts, and Aztecs, developed base 20 number systems probably from originally using both fingers and toes to count. The Sumerians and the Babylonians, for reasons lost in the mists of time, developed a base 60 number system to which we still owe our practice of dividing the hour into 60 minutes, the minute into 60 seconds, and the circle into 360°.

The most direct ancestor of our particular number system, with its rather abstract but indispensable notion of zero, and its use of the position of a given digit to determine the quantity represented by that digit, was developed in India more than a thousand years ago. During the great flourishing of Islam, it slowly spread throughout the Arab world before making its way to Europe. The great popularizer of the Hindu-Arabic number system in Europe was Leonardo of Pisa, a well-traveled mathematician whose 1202 publication *Liber Abaci* brought Europe out of the mathematical dark age of the Roman numeral.

0.2 THE DECIMAL SYSTEM: A CLOSER LOOK

If your niece shows you a jar full of pebbles that she gathered at the beach and tells you that she counted them there are exactly 608 pebbles in the jar, what exactly is the relationship between the sequence of numerals ‘‘608’’ and the number of pebbles in the jar? How do we make this connection?

The number 608 has three digit positions or numeral slots: the 100s position (which contains a six), the 10s position (which contains a zero), and the ones position (which contains an eight). Every base 10 number consists of numerals in such slots or digit positions. The total value represented by a given numeral in a number is determined by two things: the value of the numeral itself and the value assigned to the position that the numeral occupies in the number. To get the total value carried in each position, we multiply the numeral in the position by the value of the position itself. To get the value of the entire number, we add up the total values for all the positions. Thus, for the number 608,

$$6 \times 100 + 0 \times 10 + 8 \times 1 = 600 + 0 + 8 = 608.$$

Table 0.1 shows the relationship between each digit position and the value we assign it. Each position represents exactly 10 times more than the position to its right and exactly one-tenth the value of the position to its left. Another way of saying this is that the positions represent successive powers of 10. If we count the positions from right to left beginning with zero (not one), each position carries a value equal to 10 raised to the power of the position number. Stated more mathematically, each position carries the value 10^x where x is the number of the

TABLE 0.1

6	0	8
100s 10^2 Position 2	10s 10^1 Position 1	1's 10^0 Positon 0

position. For example, position 2 is the 10^2 or 100s position. Position 17 would be the 10^{17} or 100,000,000,000,000,000's position.

In any number the digit in the leftmost position is called the **most significant digit** because it carries the most value in the number. The rightmost digit is called the **least significant digit** because it carries the least value in the number.

Look at our example number 608 again. What about the 0 in the 10s position? It is tempting to say that there are no 10s in 608 but that is not really true; there are 60 tens (plus one-eighth of a 10, if you want to get picky). What the 0 in the 10s position really indicates is that there are no leftover 10s that cannot be lumped together into a hundred or a thousand, etc.

Likewise, the 6 in the 100s position means that there are six 100s left over that cannot be lumped together into a thousand (or some even higher power of 10). If the number were 908, there would be nine 100s that could not be lumped together into a thousand. However, if we were to add another hundred to 908, we would have 10 hundreds, which would be an even thousand with no 100s left over, or 1008.

Because each digit position only holds the leftover numbers that could not be lumped together into a higher power of 10, each digit only needs to hold at most nine. Nine is as full as a digit can get before it ''flips over'' into the next digit position.

The numeral 0 is somewhat different from the rest of the base 10 numerals, 1 through 9. It does not actually represent any value. It represents nothing, which is different than saying that it does not represent anything. It is sometimes called a place holder because it occupies space in a number without adding any value to it. Yet without the lowly 0, how could the 1 in 1,000,000 occupy the millions position?

Another property of 0 is that it invisibly occupies all digit positions to the left of the most significant digit of a number; we just do not write them all out for convenience's sake. Thus, 4 is the same as 0000000004. All those 0s mean that there are no 10s, no 100s, no 1000s, etc. When we do choose to write the meaningless 0s in the front of a number, they are sometimes called **padding zeros.** Sometimes padding zeros are necessary when filling out forms (e.g., month of birth: 03).

0.3 OTHER BASES

What if people had only seven fingers? There would only be seven digits: 0, 1, 2, 3, 4, 5, and 6 but no 7, 8, or 9. We would count in base 7, in which a digit ''flips over'' into the next higher position when it reaches six not nine. If you had a car with a base 7 odometer, as you drove it would climb mile after mile toward 666666, at which point it would flip back to 000000. Counting in base 7 works as shown in Table 0.2.

TABLE 0.2

Base 10	Base 7
0	0
1	1
2	2
3	3
4	4
5	5
6	6
7	10
8	11
9	12
10	13
11	14
12	15
13	16
14	20
15	21
...	...
46	64
47	65
48	66
49	100
50	101
51	102
52	103

The digit positions in base 7 do not represent the same values that they do in base 10. In base 10, each digit position represents a value 10 times as great as that represented in the position to its right. In base 7 each digit position represents a value seven times as great as that represented by the position to its right and one-seventh as great as that of the position to its left. That is, digit positions represent successively greater powers of seven. There is a ones' position at the far right, followed by a sevens position, followed by a 49s (7^2) position, followed by a 343s (7^3) position, and so forth. The values assigned to each of the first five base 7 digit positions are shown in Table 0.3.

TABLE 0.3

Position 4	Position 3	Position 2	Position 1	Position 0
7^4	7^3	7^2	7^1	7^0
2401	343	49	7	1

EXERCISE 0.1

A. Write out (in base 10) the values assigned to the first five digit positions of the following bases, beginning with the least significant digit position (zero) and working toward the most significant digit position (four):

1. 9
2. 5
3. 4
4. 3
5. 2

0.4 CONVERTING FROM BASE 7 TO BASE 10

If we have a number in base 7 e.g., 4625, how do we know how big it is? Because we are not used to working in the base 7 number system, for it to mean anything to us we must be able to convert it to base 10.

To convert a number from base 7 to base 10, we multiply the numeral in each digit position by that digit position's value then add the resulting products together. For example, to find the base 10 representation of the base 7 number 4625, we note that there is a four in the 343s position, a six in the 49s position, a two in the sevens position, and a five in the ones position, as shown in Table 0.4.

Based on this observation, we calculate:

$$4 \times 7^3 + 6 \times 7^2 + 2 \times 7^1 + 5 \times 7^0 = 4 \times 343 + 6 \times 49 + 2 \times 7 + 5 \times 1 = 1372 + 294 + 14 + 5 = 1685.$$

To avoid confusion in places where confusion is likely, we will put a subscript on numbers indicating their base. Thus, $4625_7 = 1685^{10}$. Naturally, the subscripted number itself, indicating the base, is in base 10.

Consider for a moment the quantity 25_{10}. It represents the same quantity

TABLE 0.4

4	6	2	5
7^3	7^2	7^1	7^0
343	49	7	1

that 34_7 does. To illustrate this, look at Figures 0.1 and 0.2. Each contains the same number of chili peppers, but in each group the chili peppers are arranged differently to reflect the different ways of looking at the same quantity depending on the number system being used.

Figure 0.1 represents 25 chili peppers grouped in a way that makes sense in base 10: two groups of 10 chili peppers each and one group of five chili peppers left over. Figure 0.2 shows the same number of chili peppers but grouped in a way that reflects a base 7 method of counting them: three groups of seven chili peppers each and one group of four chili peppers left over.

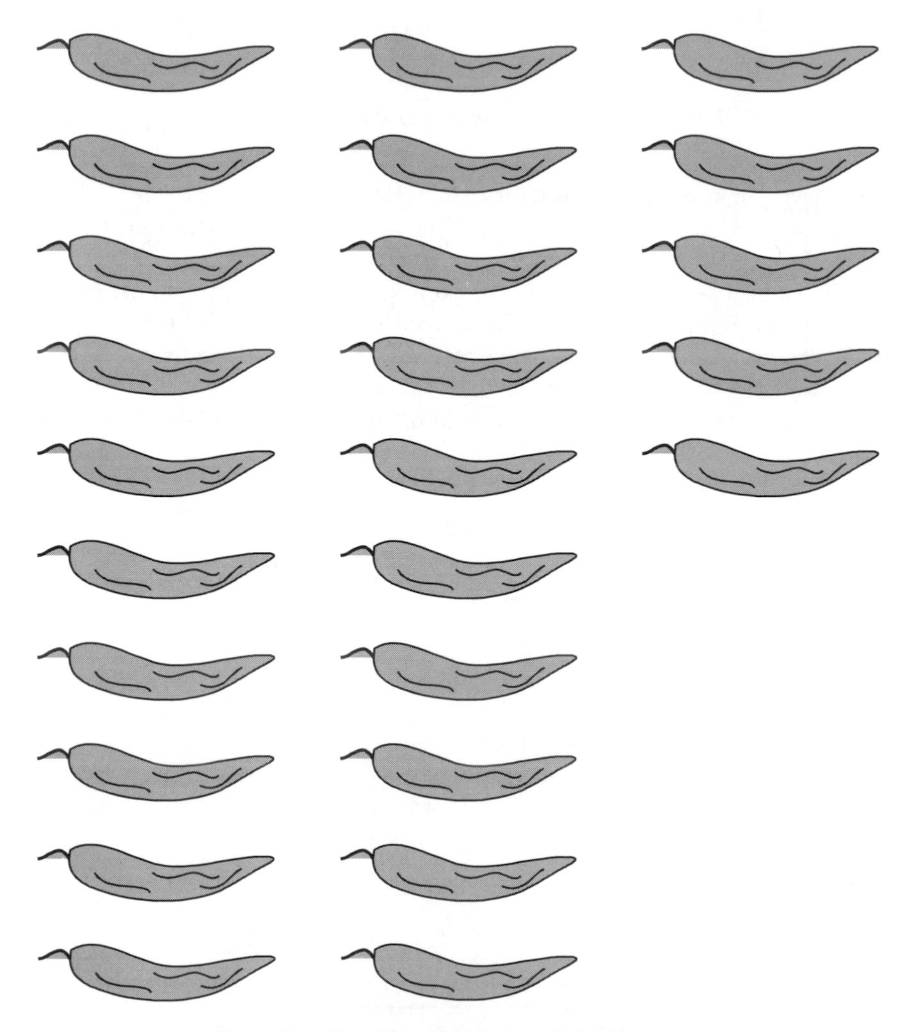

Figure 0.1 Base 10 style grouping of 25 chili peppers.

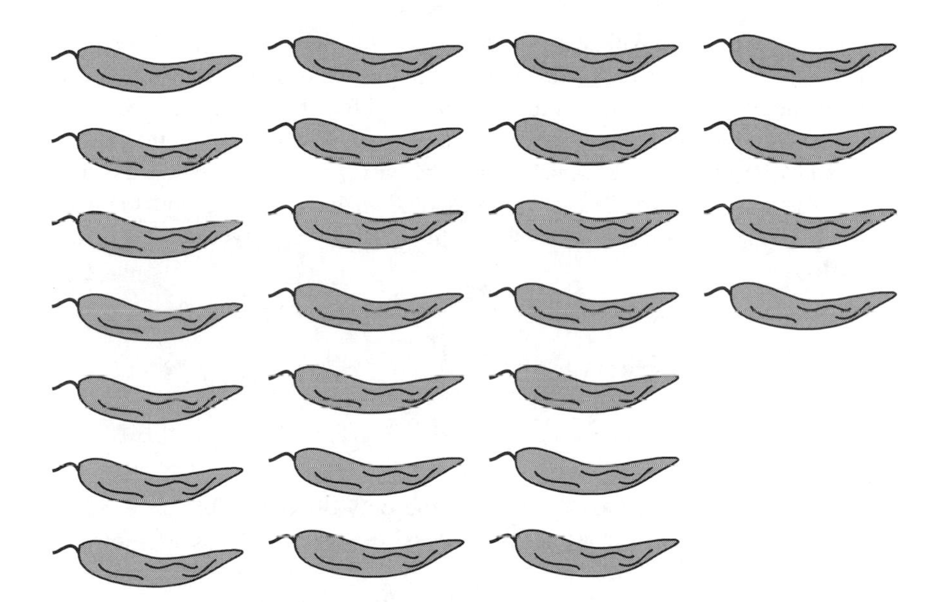

Figure 0.2 Base 7 style grouping of 25 chili peppers.

0.5 CONVERTING FROM BASE 10 TO BASE 7

To convert a number from base 10, which we are used to, to another base (e.g., base 7), we need to rearrange the number from groupings of powers of 10 to powers of 7. We do not draw out chili peppers and move them around as we did in the last example. We regroup the number mathematically using division.

Recall that each digit in a base 7 number can be thought of as ''leftover,'' quantities that were left over after forming groups of greater powers of seven. This concept of ''leftovers'' is the same as the concept of **remainders** in long division. To find the base 7 representation of a number whose base 10 representation is known to us, we keep dividing the number by 7 according to the normal rules of long division. The remainder that results from each division gives us one more digit in our base 7 number, working right to left, from the least significant digit to the most significant digit.

For example, to find out what 967_{10} is in base 7, first we divide 967 by seven: $967 \div 7 = 138$, with a remainder of one. (If we had 967 ball bearings and we tried to form them on the ground in neat groups of seven, there would be one left over.) Thus, there is a one in the ones column of our base 7 number. Next, we take the quotient we got before, 138, and divide it by 7. This is equivalent to grouping the groups of seven ball bearings we had before into groups of seven, making groups of 49 ball bearings each: $138 \div 7 = 19$, with a remainder of five. Thus, there are five groups of seven ball bearings that do not fit evenly into any of the groups of 49 ball bearings. Therefore, the digit in the sevens position in the base 7 number is five. To find the digit in the 49s position, we again divide our quotient by 7: $19 \div 7 = 2$, with a remainder of five, therefore a five occupies the 49s position in our base 7 number. Finally, $2 \div 7 = 0$ with a remainder of two, so a two occupies the 343s position in the base 7 number. Now we are done because our number has melted away to zero, and we have nothing more to divide by seven: $967_{10} = 2551_7$. We can check this result by reconverting 2551_7 into base 10:

$$2 \times 7^3 + 5 \times 7^2 + 5 \times 7^1 + 1 \times 7^0 =$$

$$2 \times 343 + 5 \times 49 + 5 \times 7 + 1 \times 1 =$$

$$686 + 245 + 35 + 1 =$$

$$967.$$

Now let us consider base 5, in which the only digits are 0, 1, 2, 3, and 4 and the digit positions have values of successive powers of five, as shown in Table 0.5.

To convert a base 5 number to base 10, we follow the same method we used for base 7: multiply each numeral by the value its digit position represents. For example, to convert 413_5 to base 10, we calculate as follows:

$$(4 \times 5^2) + (1 \times 5^1) + (3 \times 5^0) =$$

$$4 \times 25 + 1 \times 5 + 3 \times 1 =$$

$$100 + 5 + 3 = 108_{10}.$$

To convert back to base 5, we use the same sequence of steps for converting base 10 to base 7:

$$108_{10} \div 5 = 21, \text{ remainder } 3 \qquad \text{(3 in 1's position)}$$

$$21_{10} \div 5 = 4, \text{ remainder } 1 \qquad \text{(1 in 5's position)}$$

$$4 \div 5 = 0, \text{ remainder } 4 \qquad \text{(4 in 25s position)}$$

The two methods for converting from base 10 to another base and back

TABLE 0.5

Position 5	Position 4	Position 3	Position 2	Position 1	Position 0
5^5	5^4	5^3	5^2	5^1	5^0
3125	625	125	25	5	1

again are general and can be applied to any base. Such a distinct sequence of steps is called an **algorithm.**

What is the smallest base we could have? Because any number system has digits 0 through one less than the base of the number system itself, we could not really have a base 1 because it would only have one digit, 0. With only one symbol, we could not use our method of using the digit position to impart value to a numeral because we would have no place holder. However, we can have a base 2 with digits 0 and 1. The base 2 or **binary** number system would be extremely unwieldy to use when balancing your checkbook, for example, but it has some interesting properties that make it of crucial concern for the rest of this book.

On the other hand, could we have a base greater than 10? We could, but only if we invent new digits to represent single-digit numbers greater than nine. By convention, people who need to use such number systems (and there are more of them than you might think) usually use letters to represent these digits. Accordingly, counting in base 16 works as shown in Table 0.6.

The base 16 or **hexadecimal** number system has 16_{10} digits: 0 through F ($F_{16} = 15_{10}$).

TABLE 0.6

Base 10	Base 16
0	0
1	1
2	2
3	3
4	4
5	5
6	6
7	7
8	8
9	9
10	A
11	B
12	C
13	D
14	E
15	F
16	10
17	11
18	12
19	13

EXERCISE 0.2

A. Count from one to 20_{10} in the following bases:

1. 8
2. 6
3. 5
4. 3
5. 2

B. Convert the following numbers to their base 10 equivalents:

6. 3124_7
7. 12332_4
8. 15366_9
9. 4000_5
10. 5023_8
11. 77_8
12. 333_4
13. 111111_2

C. Convert the following base 10 numbers to the indicated base:

14. 413_{10} to base 5
15. 128_{10} to base 8
16. 3000_{10} to base 6
17. 963_{10} to base 3
18. 67_{10} to base 2

D. Perform the indicated base conversions:

19. 54_8 to base 5
20. 312_4 to base 7
21. 516_6 to base 7
22. 12212_3 to base 9
23. 100_8 to base 2
24. What generalizations can you draw about converting a number from one base to a power of that base, e.g., from base 3 to base 9 (3^2) or from base 2 to base 4 (2^2) or base 8 (2^3)?

0.6 ADDITION IN OTHER BASES

Adding numbers in other bases is much like adding numbers in base 10. Care must be taken, however, because it is easy to forget that you are working in another base. For instance, in base 6, 4 + 3 does not equal 7, as it does in base 10 because there is no numeral 7 in the base 6 number system. In base 6, 4 + 3 = 11; a one in the sixes position and a one in the ones position, $11_6 = 7_{10}$.

What about adding multidigit numbers? The familiar rules of carrying apply. To perform the following addition in base 4:

$$\begin{array}{r} 32113 \\ +11323 \\ \hline \end{array}$$

We begin with the first column and add the numbers in it together: $3_4 + 3_4 = 12_4$. We would then drop the 2 into the first column of the sum and carry the 1:

$$\begin{array}{r} 1 \\ 32113 \\ +11323 \\ \hline 2 \end{array}$$

Next, we add the numbers in the second column, including the carry: $1_4 + 1_4 + 2_4 = 10_4$. We drop the 0 and carry the 1. We continue in this vein until we have

$$\begin{array}{r} 1111 \\ 32113 \\ +11323 \\ \hline 110102 \end{array}$$

We can check that $32113_4 + 11323_4 = 110102_4$ by converting all three numbers to base 10 performing the addition on the first two numbers in base 10 and making sure that the sum we get is equal to the third number.

First number:

$32113_4 = 3 \times 4^4 + 2 \times 4^3 + 1 \times 4^2 + 1 \times 4^1 + 3 \times 4^0 =$

$3 \times 256 + 2 \times 64 + 1 \times 16 + 1 \times 4 + 3 \times 1 =$

$768 + 128 + 16 + 4 + 3 =$

919_{10}.

Second number:

$11323_4 =$

$1 \times 4^4 + 1 \times 4^3 = 3 \times 4^2 + 2 \times 4^1 + 3 \times 4^0 =$

$1 \times 256 + 1 \times 64 + 3 \times 16 + 2 \times 4 + 3 \times 1 =$

$256 + 64 + 48 + 8 + 3 =$

379_{10}.

Sum:

$110102_4 =$

$1 \times 4^5 + 1 \times 4^4 + 0 \times 4^3$

$+ 1 \times 4^2 + 0 \times 4^1 + 2 \times 4^0 =$

$1 \times 1024 + 1 \times 256 + 0 \times 64 + 1 \times 16 + 0 \times 4 + 2 \times 1 =$

$1024 + 256 + 0 + 16 + 0 + 2 =$

1298_{10}.

$919_{10} + 379_{10} = 1298_{10}$.

EXERCISE 0.3

A. Perform the following additions in the indicated bases. Check your work by converting each number added to base 10, performing the addition, then converting the resulting base 10 sum back into the indicated base.

1. $16_8 + 5_8$
2. $24_5 + 132_5$
3. $524_6 + 312_6$
4. $713_9 + 238_9$
5. $442_6 + 115_6$
6. $265_6 + 333_6$

0.7 COUNTING

Combinatorics is the science of counting different combinations of things. We are interested in counting combinations of numerals. In this pursuit, we think of numerals as symbols only, distinguishable from one another, but not necessarily as numbers representing quantities. Given this understanding of numerals, 00 and 03 are two different two-digit numbers. How many different two-digit base 10 numbers are there? When we list them all from the smallest, 00, to the largest, 99, we see that the numbers 01 to 99 count themselves. There are 99 numbers from 01 to 99, plus one more for 00. Thus there are exactly 100, or 10^2, two-digit base 10 numbers.

Similarly, there are a million (10^6) six-digit base 10 numbers, 000000 through 999999. That is, there are 999999 numbers from 000001 to 999999, plus one more for 000000. For any base, to find how many different numbers there are in that base of a certain number of digits, we raise the number of the base to the power of the number of digits. Put more mathematically, there are always x^n different n-digit base x numbers.

How many different five-digit base 4 numbers are there? Our formula says there should be 4^5, or 1024_{10}, or 100000_4 of them. If we were to list them from the smallest five-digit base 4 number to the largest, the list would look like this: 00000, 00001, 00002, . . . , 33331, 33332, 33333. As before with the base 10 numbers, the numbers from 00001 to 33333 count themselves, so there are 33333_4 of them. Adding one more for the digit combination 00000 yields $33333_4 + 1_4 = 100000_4 = 1024_{10}$ different five-digit base 4 numbers. This means, in general, that there are 1024_{10} ways of arranging five of any symbol that can take on four distinct values.

Imagine that you have a large bag of jelly beans. There are six different colors of jelly beans: green, red, blue, brown, tan, and orange. If, without looking, you reach into the bag and take out three jelly beans and lay them out in a row, how many different combinations of colors could you have? Since there are six different colors, we can imagine that each color represents a different base 6 numeral (0–5), and our problem becomes the same as asking how many different three-digit base 6 numbers there are. Thus there are $6^3 = 1000_6 = 216_{10}$ possible different color combinations in a row of three randomly chosen jelly beans.

As you might imagine, combinatorics is a branch of mathematics that has powerful applications. It is closely related to the mathematical field called **statistics,** which deals with probabilities of given events occurring or not occurring. Card sharks in Las Vegas and Atlantic City train themselves in the techniques of combinatoric analysis to determine the likelihood of certain combinations of cards appearing in their hands and in those of their opponents across the table.

EXERCISE 0.4

1. How many four digit base 5 numbers are there? Give your answer in base 10 and in base 5.
2. Plot a chart with the numbers 2 to 10 on the horizontal axis (x), and 0 to 100 in units of 10 along the vertical axis (y). For each value of x (2 through 10) plot a point showing how many two-digit base x numbers there are. Connect the ten resulting points.
3. Why did the x values in the previous exercise begin with two and not zero or one?
4. If seven people in a row on an airplane are each given the choice of coffee, tea, or milk by the flight attendant, how many different combinations of people and drinks could there be in that row of people, assuming each person has something to drink?
5. How many combinations of people and drinks could there be in the situation described in the previous exercise if we consider the possibility that any of the seven people in the row could have declined the offer of a drink?
6. How many combinations of people and drinks could there be (including the possibility that any passenger may not drink at all) on the entire plane if there are 213 passengers? (Do not write out the entire resulting number—represent it using exponents).

7. When we **shift** a number one position to the right, we move each digit in the number one place to the right, allowing the least significant digit to disappear off the end. Thus 6377 shifted one place to the right becomes 637. Similarly, when we shift a number to the left, we move each digit in the number to the left, appending zeros to the least significant digit as needed. Thus, 6377 shifted one place to the left becomes 63770. When a base 10 number is shifted to the left, what implications does that have for the actual quantity represented by the number? What happens to a base 10 number when it is shifted to the right?
8. Give the base 10 representation of 422_5. Shift 422_5 to the left one place and two places, and to the right one and two places. Give the base 10 representation of all four resulting numbers. Characterize the effect of shifting a base 5 number to the left or right.
9. Shift 11010010_2 to the right, converting to base 10 after each shift, until there is nothing left. What is the relationship between the base 10 numbers in the resulting sequence?
10. In general, how does shifting a base x number y places to the right or left affect the quantity represented?

0.8 THE BINARY NUMBER SYSTEM

Of particular interest is the base 2 or binary number system. In the binary system, there are only two digits, 0 and 1. Binary digits are called **bits** (for **b**inary dig**it**). The value of each digit position in a binary number is exactly double that of the position to its right and exactly one-half that of the position to its left.

Binary numbers get big quickly as one counts. For example, $256_{10} = 2^8 = 100000000_2$. Counting in base 2 is shown in Table 0.7 (for a longer list see Appendix A).

TABLE 0.7

Base 10	Base 2
0	0
1	1
2	10
3	11
4	100
5	101
6	110
7	111
8	1000
9	1001
10	1010

TABLE 0.8

128	64	32	16	8	4	2	1
2^7	2^6	2^5	2^4	2^3	2^2	2^1	2^0

Listed in Table 0.8 are the first eight powers of two in base 10. For a larger list of the powers of two, see Appendix B.

In all of the other bases we have seen, a numeral in a particular digit position told us how many of that digit position's units there were that could not be lumped into a higher power of that base. The 3 in the fives (5^1) digit position in 42231_5 tells us that there are three leftover fives that can not be lumped together into a 5^2 or a higher power of five.

In the binary number system, however, there are only zeros and ones. So the question of how many of a particular power of two there are that cannot be lumped together into a higher power of two has two possible answers: zero (none) or one. Thus we can rephrase the question from: "*How many* fours (2^2) are there in this number that cannot be lumped into an eight (2^3) or a higher power of two?" to "*Is there or is there not* a four in this number that can be lumped together into an eight or a higher power of two?" If there is a one in the fours digit position (as there is in the binary number 10010100), the answer is yes, there is a four that cannot be lumped into a higher power of two. If there is a zero in that position (as in the number 1101001001) the answer is no.

This is why the binary system is of particular interest to us. It is more than a number system that tells us "how many?" although it works perfectly well as such. It is also a logic system that tells us "is there?" Bits are more than numerals. They can represent **numerical quantities,** as numerals in other bases can, but they also represent **logical qualities.** A 0 and a 1 can mean the numerals 0 and 1, or 0 can mean false and 1 can mean true. 0 can mean off and 1 can mean on. The interpretations we may assign to the symbols 0 and 1 are endless.

Using the techniques we learned previously, we can see that the largest possible eight-bit number is 11111111, the smallest is 00000000, and there are $100000000_2 = 2^8 = 256_{10}$ different eight-bit binary numbers (00000001 through 11111111 (255_{10}) plus one more for 00000000 makes exactly 100000000_2 eight-bit numbers in all).

Similarly, there are 2^2 or four different two-bit numbers. They range from 00 to 11. There are few enough of them that we can list them: 00, 01, 10, 11. These four represent every possible combination of two bits.

Given two bits, let the first bit represent the color of your shoes, brown = 0 or black = 1, and the second bit represent the pattern on your socks, plain = 0, argyle = 1. Given these choices, how many different combinations of shoes and socks could you wear? We know that the answer is four, because we have two choices of two options each, or two bits. Again, we could list the four combinations: brown shoes and plain socks (00), brown shoes and argyle socks (01), black shoes and plain socks (10), or black shoes and argyle socks (11).

Given a choice of black or brown shoes, plain or argyle socks, and a white or pink shirt, how many total combinations are there? We have added another bit (bit 1: white shirt = 0, pink shirt = 1; bit 2: brown shoes = 0, black shoes = 1; bit 3: plain socks = 0, argyles = 1). We already know that three bits can range from 000 to 111 and that there are 2^3 or eight possible combinations: 000, 001, 010, 011, 100, 101, 110, 111. Thus, there are eight different combinations of shoes, socks, and shirts. Look at the list of combinations of three bits again. If you ignore the first bit, the last two bits seem to cycle through the same four combinations twice—once with the first bit off, then again with the first bit on. By adding the complication of the choice of shirts to the problem of shoes and socks, we doubled the number of combinations because now we consider all of the original four combinations of shoes and socks twice: once with a white shirt, and again with a pink shirt.

0.9 COMBINATORIC EXAMPLES

The following examples illustrate the power of the binary number system with regard to combinatoric analysis of a variety of situations.

0.9.1 U.S. Presidential Election Example

Given that in a presidential election, each state in United States votes Democrat or Republican, how many different ways could all 50 states vote (ignoring for the moment such complications as independent candidates or the peculiarities of the electoral college)? That is, from a Democrat shutout in which all 50 states voted Democrat, to a Republican shutout and all combinations in between of different states voting Democrat and Republican, how many different combinations of voting patterns are there for all 50 states taken together?

Assign one bit to each state in order, and let each state's bit be 0 if that state votes Democrat, and 1 if it votes Republican. Then the entire country may be thought of as one big 50-bit binary number, and there are 2^{50} (an almost impossibly huge number) different combinations of state-level outcomes in a national election.

0.9.2 Pizza Example

If you were at a pizza parlor that sold one size of pizza and seven toppings were available (mushroom, sausage, pepperoni, onion, pepper, anchovies, and extra cheese), how many different kinds of pizza could you order?

Imagine one bit for each topping, seven bits in all. We can represent any particular kind of pizza by a seven-bit binary number, in which each topping's bit tells us whether that topping is present on the pizza (1 = topping is on the pizza, 0 = topping is off the pizza). Thus, all seven bits off (0000000) represents a plain cheese pizza with nothing on it, while all seven bits on (1111111) represents a pizza with the works. It should be clear that there are as many different kinds of pizza as there are combinations of seven bits, or $2^7 = 10000000_2 = 128_{10}$ (see Table 0.9).

TABLE 0.9 This Combination of Topping Bits Represents a Pizza with Mushroom, Pepperoni, and Extra Cheese

1	0	1	0	0	0	1
mushroom	sausage	pepperoni	onion	pepper	anchovy	extra cheese

0.9.3 Hypercube Example

As a less mundane example, consider the mind-bending hypercube. A hypercube is an abstract mathematical construct which is essentially a square or cube that exists in an arbitrary number of dimensions. To begin with, look at the line segment of length 1 on the number line shown in Figure 0.3.

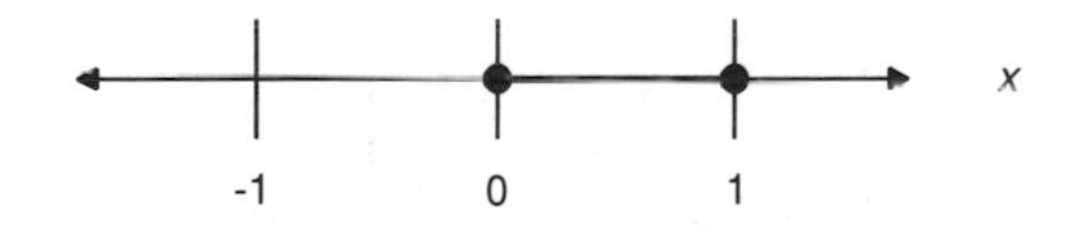

Figure 0.3 Line segment (one-dimensional hypercube).

This line exists in only one dimension. It has length, but no width or depth. It is considered a one-dimensional hypercube. It has two end points, at 0 and 1, so to specify any endpoint on the hypercube completely, we need only specify one bit (let us call it x). Its end points are called $x = 0$ and $x = 1$.

Look at the square in Figure 0.4, graphed on a Cartesian grid. Each of its sides is of length 1. This is a two-dimensional hypercube. It has length and width, but no depth. There are four endpoints (only now we will call them **vertices** or **corners**). To specify a particular vertex we need two bits. The four vertices are shown on the graph and are at the points $(x, y) = (0, 0)$, $(x, y) = (0, 1)$, $(x, y) = (1, 0)$, and $(x, y) = (1, 1)$. Note that the four vertices comprise all possible combinations of two bits, just as the two endpoints on the one-dimensional hypercube ($x = 0$ and $x = 1$) comprised all possible combinations of one bit.

Now look at the hypercube in three dimensions in Figure 0.5, which we know as the cube. This figure has eight vertices, and each requires three bits to specify it completely.

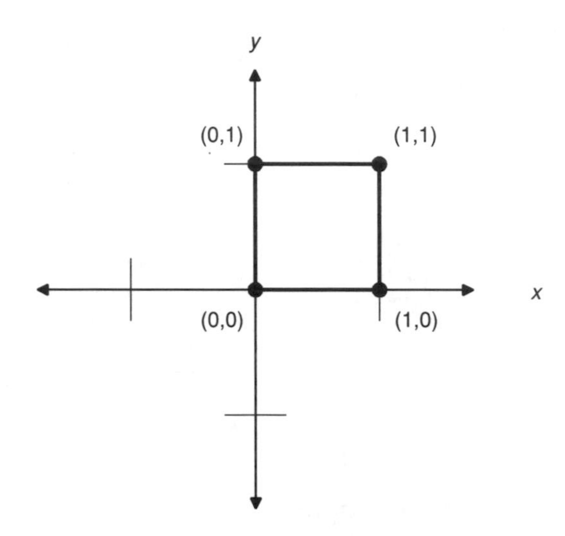

Figure 0.4 Square (two-dimensional hypercube).

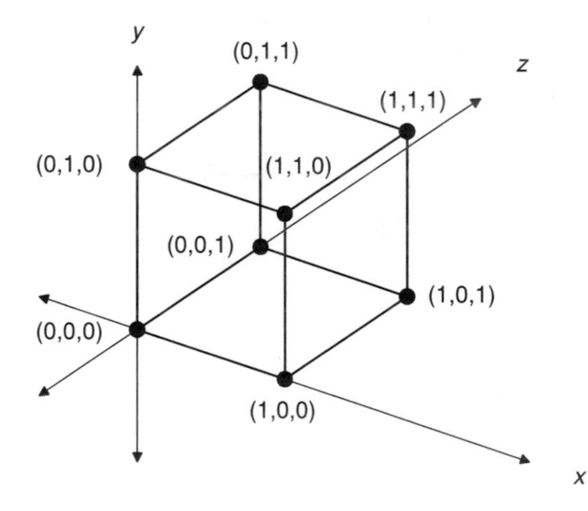

Figure 0.5 Cube (three-dimensional hypercube).

Based on these examples, we may make some generalizations about hypercubes in any dimension. While most sane people have a hard time visualizing a four-dimensional object, it is safe to say that if there was such a thing as four-dimensional space, a hypercube in that space would have 16 vertices. Each of these vertices would require four bits to specify it completely, one bit for the coordinate along each axis on its graph. In general, for any number n, in n dimensions a hypercube has 2^n vertices, each of which requires n bits to specify completely. These results are summarized in Table 0.10.

TABLE 0.10

Number of Dimensions	Number of Vertices	Number of Bits Needed to Specify a Vertex
1	2	1
2	4	2
3	8	3
4	16	4
n	2^n	n

Far from being just a brain toy for mathematics professors, the hypercube in dimensions greater than three has some useful real-world applications, even though it does not actually exist. For instance, in some supercomputers it has been found that an effective way to connect different computing components is as if they were vertices on a hypercube and the communications channels between them were the sides of the hypercube. This way, the efficiency of communication between the large number of computing components is maximized.

0.9.4 Binary Trees

A **binary tree** (Fig. 0.6) a diagram used in a number of disciplines. It consists of circles, dots, or boxes called **nodes** and lines called **branches** that connect the nodes. It starts at the top with a single **root node,** which is at level zero. Two branches descend from the root node, one to the left, and one to the right. Each of these branches leads to another node. Each of these two nodes is at level one. They are said to be children of the root node; the root node is their parent. Throughout the binary tree, each node is branched to by a node in the level above it (its parent) and has branches to two nodes in the level below it (its children). A binary tree could go on forever, but generally they end sometime with child

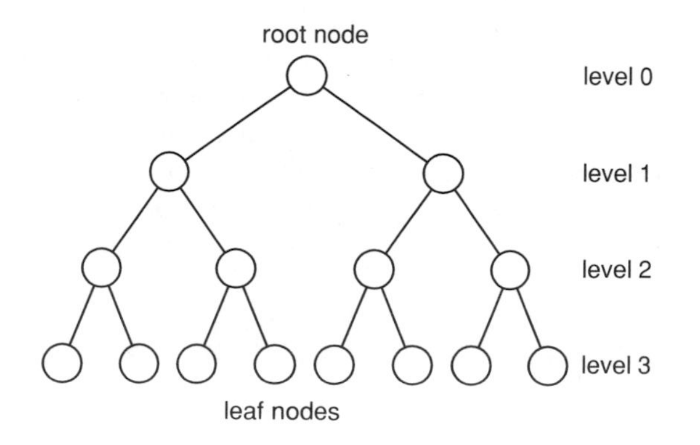

Figure 0.6 A four-level binary tree.

nodes that have no children of their own. These nodes are called **leaves** or **leaf nodes.**

Binary trees have several interesting properties. Each level has exactly twice the number of nodes as the level above it. Beginning with the root node at level zero, level n has 2^n nodes in it, and the entire tree above level n (i.e., levels 0 through $n - 1$) has $2^n - 1$ nodes in it or one less node that level n has. For example, letting n in the previous sentence be 13, level 13 alone has 2^{13} nodes in it (8192), and there are $2^{13} - 1$ (8191) nodes in the entire tree in rows 0 through 12.

The terms ''child'' and ''parent'' suggest a genealogical table, although a binary tree only describes successive generations if each person (node) has exactly two children. However, if we think of parents in the binary tree as human children and vice versa, (conceptually turning the tree upside down) a binary tree exactly describes your relationship to your ancestors. If you are the root node, your parents are the two level-one nodes, your grandparents are the four level-two nodes, etc. How many great-great-great-great-great-great-great-great grandparents do you have? There are eight ''great''s there, so that means these are the level 10 ancestors on the binary tree. Thus are $2^{10} = 1024_{10}$ of them.

Binary trees are also sometimes called **decision trees.** Each node can be thought of as a yes/no decision. A yes (1) makes you take the branch to the right. A no (0) makes you take the branch to the left. At each level, you make a choice, starting at the root node. By the time you get to level n, you have made n decisions, and whatever node you are on can only be reached by making the exact sequence of choices that you made. Thus, starting at the root node and making only eight yes/no decisions, you find yourself at level eight, at one node among 255 others at that level, each of which could have been reached with a different combination of eight yes/no choices.

If you wrote down a 0 or a 1 reflecting the choice you made at each of the eight levels (0–7) as you ''walked'' a binary tree in this way down to level eight, you would have an eight-bit binary number that uniquely represented the particular sequence of choices that you had made, as well as which uniquely labeled the particular level-eight node you had reached by that sequence of choices. Anyone who has played the game 20 Questions will realize that virtually anything in the world may be singled out on the basis of 20 yes/no answers. Theoretically, then, everything in the world could be given a unique 20-bit identification number, allowing for 2^{20} things in the world (about a million). This, however, ignores the fact that the questions asked in the game of 20 questions change on the basis of previous answers.

Tennis tournament charts are arranged in binary tree form. All players start out as leaves at the bottom of the tree and only move up to the level above if they win their match against their ''sibling'' node. Each round of matches eliminates half of the players. Finally, only the player who wins all her matches gets to the root node. Since a binary tree with eight levels of decisions in it (a nine-level tree) has 256 leaves, it takes only eight rounds of matches to whittle 256 (2^8) players down to the one champion.

EXERCISE 0.5

1. How many combinations of heads and tails are possible if a coin is flipped seven times? How many combinations if it is flipped n times?
2. If a given radioactive substance has a half-life of one year, then each year that passes leaves the substance half as radioactive as it had been a year before and twice as radioactive as it will be one year hence. Rounded up to the nearest year, after how many years would such a substance be one three-hundredth as radioactive as it was to begin with? How long would it take for such a substance to lose all of its radioactivity?
3. Arman and Mike review movies on television. Each of them gives each of the five movies reviewed a thumbs up or a thumbs down. On a particular show, how many different ways could the reviews turn out?
4. Describe as well as you can a zero-dimension hypercube.
5. How many combinations of options are possible on a car that has 37 optional features?

1

The Basic Functions of Boolean Algebra: AND, OR, and NOT

Boolean algebra is a system of mathematics (or a system of logic, depending on how you look at it) that, while simple in its operations, has an extremely rich diversity of applications and interpretations. Technically the term **Boolean algebra** refers to a family of such mathematic or logic systems, but it is almost always understood to mean one system in particular. This book is about this most common form of Boolean algebra. Hereafter, our use of the term ''Boolean algebra'' will refer exclusively to this type of Boolean algebra.

Before we begin our study of Boolean algebra in detail, it is a good idea to delve into its history. In the western world for many centuries the subject of logic as a field of study was not only dominated by the Greek philosopher Aristotle (384–322 B.C.), it began and ended with him. For more than two thousand years, as Immanuel Kant said, logic neither advanced nor retreated a single important step.

In Victorian England, however, several mathematicians and philosophers began to improve upon Aristotle, chiefly by using symbols instead of complete sentences to represent logical propositions and manipulating those symbols according to certain explicit rules. This abstraction of the whole process of logical inference blurred the line between logic and mathematics and gave rise to the new field of **symbolic logic.**

George Boole (1815–1864) is arguably the inventor of this field, as his system of symbolic logic was the most completely realized for its time. While others were working in the same direction, Boole made the first leap to a truly symbolic system of abstract symbols resting upon formally stated axioms rather than using various forms of symbol operations to illustrate or supplement a more traditional (and verbal) approach to logic.

George Boole was born into modest means, the son of an academically inclined cobbler. He was largely self-educated (there was little money for formal schooling) and became a teacher at an early age. In 1855 he married Mary Everest, niece of Sir George Everest (for whom Mount Everest is named). In 1864 he died from pneumonia he contracted after walking to class in the rain then lecturing in wet clothes.

Boole's most important work is rather immodestly titled *An Investigation of the Laws of Thought On Which Are Founded the Mathematical Theories of Logic and Probabilities.* Boole, like many of his contemporaries,[1] naively thought that all of the functions of the human mind could be rendered in the form of a rigidly precise logical system with its workings made explicit by the formal rules of that system. More than a century of symbolic logic and several decades of artificial intelligence research have made it clear that the principles according to which the human mind works (if there are any) are substantially more ''squishy'' than George Boole would have dreamed, and whatever intelligence is, logic, as conceived by the Victorian philosophers, is a very narrow part of it.

While Boole's work never did reveal the secrets of the mind, it did provide a novel and interesting system of logic. It was refined and extended by later logicians: different notational conventions were adopted, new operations were added, some of Boole's operations were changed, and some were dropped. Nonetheless, what we know as Boolean algebra today remains at its core Boole's system.

Boolean algebra concerns zeros and ones and their manipulation using operations that we call **Boolean functions.** In chapter 0, we discussed bits and the binary number system, but we did so primarily from an arithmetic perspective. That is, we looked at base 2 numbers as quantities that we could count and add. However, we also introduced the concept of a bit as a logical quality, something that could be either on or off, true or false, present or not present. This more abstract logical understanding of zero and one is fundamental to Boolean algebra. For the rest of this book we shall elaborate on this idea.

Perhaps you have heard or read that computers count in base 2. This is true; computers operate on the principles of Boolean algebra. Inside the circuits of a digital computer, one voltage level is taken to mean zero and another voltage level is taken to mean one. There are different electronic devices that read and modify these voltage levels and thereby mimic Boolean functions. When used by engineers in digital devices, Boolean algebra sometimes is known as **switching**

[1] One of the contemporaries was Lewis Carroll, author of *Alice in Wonderland* and a distinguished mathematician and logician. Much of Carroll's work in symbolic logic was thought lost forever but has been recently recovered and published under the title *Lewis Carroll's Symbolic Logic.*

algebra because such devices are thought of as complex networks of switches that switch between zero and one.

Several of Boole's contemporaries thought about creating machines that would implement Boolean functions. William Stanley Jevons built a "logical piano" for this purpose no later than 1869. In a long lost but recently discovered letter he wrote in 1886 to a mechanically inclined friend, the American philosopher and logician Charles Sanders Peirce (1839–1914) suggested using electricity to implement Boolean functions, going so far as to sketch the earliest known circuits based on Boolean algebra. For better or worse, Pierce's suggestion was lost in the sands of history. In 1910 the idea resurfaced in a review of Louis Couturat's, "Algebra of Logic" by a Russian physicist, Paul Ehrenfest. In his paper, Ehrenfest mentioned in passing that Boolean algebra could perhaps be used to create automatic electrical telephone switches. Oddly, his suggestion was ignored for almost 30 years, another historical near miss. Finally, several people independently developed the idea of applying Boolean algebra to electrical circuits in the late 1930s[2] and the digital age began.

Computers have been made of many different types of components. The earliest direct ancestors of today's computers were electro-mechanical switches in the telephone company's vast switching banks, thus vindicating Ehrenfest's suggestion. Throughout the 1940s and 1950s, computers were made of bulky, expensive, error-prone vacuum tubes. Later they were made of individual transistors wired together. Today they are made out of chips composed of metals called semiconductors such as silicon and gallium arsenide. There are plans for computers in which the bits 0 and 1 will be represented by different intensities or wavelengths of light, and in which the "circuitry" will be special mirrors, lasers, and optical fiber. In spite of all of this diversity of physical implementation, the fundamental principles remain the same. All of these computers are based on the logic of Boolean algebra, have one kind of signal that means "0" and one kind of signal that means "1," and contain devices that operate on these signals according to the Boolean logic functions.

When we look in the chapters ahead at computers and digital circuits as examples of the principles involved, we shall see that digital circuitry is by no means the only application of Boolean algebra. George Boole certainly never imagined that his philosophical system would ever find a home in the world of electrical engineering.

1.1 BOOLEAN FUNCTIONS

In prior studies in mathematics, you have undoubtedly worked with things called numbers (whether integers, rational numbers, real numbers, or complex numbers) and functions or operations that may be performed on those numbers to produce

[2] Principal credit is usually given to Claude Shannon, whose *Symbolic Analysis of Relay and Switching Circuits* appeared in 1938.

different numbers. Examples of such functions are $+$, $\div$, $\times$, $\sqrt{}$. Perhaps you have some exposure to Euclidean geometry, in which you worked with such things as lines, points, and angles, which were operated on by functions such as bisection, rotation, or angular addition. The field of mathematics consists of the invention, manipulation, and study of such **mathematical systems** that consist of *things* and *functions* that operate on those things.

We shall explore a mathematical system called Boolean algebra. In Boolean algebra there are things called bits (0 and 1) and several functions that operate on bits to produce different bits. The bit or bits a Boolean function operates on are called its **input bits,** and the bit produced as a result of the application of the function to the inputs bits is called the **output bit.** In general, Boolean functions are not complicated, and there are only three that we need to worry about now. They are called AND, OR, and NOT. We use capital letters to distinguish the Boolean functions from the English words *and, or,* and *not.*

1.2 AND

AND takes two input bits and produces one output bit. In this regard it is similar to several of the Boolean functions that we will study. The operation of the AND function is simple. The output bit of the AND function is a 1 if and only if *both* of the input bits are 1; it is a 0 otherwise. For example, 1 AND 0 = 0, but 1 AND 1 = 1.

Just as there is a symbol, $+$, which means "plus" in the mathematical system we call ordinary arithmetic, $\wedge$ is the symbol for the AND function. Such a symbol is called an **algebraic symbol.** We use it when we write equations and other **algebraic expressions.** We will use the word AND only in English text, as the symbol $\wedge$ is easier to work with algebraically. All of the Boolean functions we will study have algebraic symbols as well as English names.

1.2.1 Logical Interpretation of Bits

In ordinary arithmetic, each function may be thought to ask a question. $x + y = z$ asks "How many is x added to y?" where x, y, and z are variables that stand for different numbers. The answer to the question is given in the form of the number represented by the variable z, the output of the function. In Boolean algebra, $x \wedge y = z$ asks "Are both x and y equal to 1?" where x, y, and z are variables that stand for bits, and 0 is understood to mean no and 1 to mean yes. The answer to the question is given in the form of the yes/no output bit represented by the variable z. This logical interpretation of bits as meaning yes or no causes the Boolean AND function to behave a lot like the English word "and."

1 = true statement: "The sun is hot."

0 = false statement: "All sheep have three legs."

To say "The sun is hot, AND all sheep have three legs" is a false statement (0) even though one of the two statements ANDed together ("The sun is hot") is true. Both statements ANDed together must be true if the resulting statement is to be true. "The sun is hot AND the Red Sox will probably lose again this year" is a true statement (1) because both of the statements that got ANDed together are true individually.

Statements of this type that must be either true or false are called **logical propositions.** Boolean algebra originally was developed to work with and manipulate such propositions.

1.2.2 Truth Table for AND

To specify the AND function completely, let us make a table for it, like a multiplication table, that shows what the result of the AND function will be on all possible combinations of inputs. The table has three vertical columns: one for each input bit, and one for the output bit. The bit in the output column on each row shows the results of ANDing both input bits on that row together.

We begin the table by writing the names of the input bits, (call them x and y), over what will become the **input columns** of the table. To the right of these headings, we write $x \wedge y$, the heading of the **output column.** We list the different combinations of the two input bits, each in a different row, beneath the input column headings. Recall from the last chapter that there are 2^2 or four possible combinations of the two input bits: 00, 01, 10, and 11. These four pairs represent the only values that the input bits for an AND function can have, so the table has only four rows (Table 1.1).

TABLE 1.1

x	y	$x \wedge y$
0	0	?
0	1	?
1	0	?
1	1	?

Now all we have to do is fill in the output column on each of the four rows with the result of ANDing the x bit and the y bit on that row together. Because

$$0 \wedge 0 = 0,$$

$$0 \wedge 1 = 0,$$

$$1 \wedge 0 = 0,$$

$$1 \wedge 1 = 1.$$

Our finished table looks like Table 1.2.

TABLE 1.2

x	y	$x \wedge y$
0	0	0
0	1	0
1	0	0
1	1	1

This table is called a **truth table.** It tells us exactly and completely what the AND function does because it shows the result of the AND function on every possible pair of bits it could be given as input. You will never know anything about the AND function that is not shown in this table. However, the table is unsurprising as it does not tell us anything not already implied by our intuitive, English language understanding of the function AND. The only 1 in the output column is in the last row, the one in which both input bits, x and y, are 1. The other three bits in the output column are 0.

Before we continue, we should establish some conventions.

1.2.3 Numbering of Rows in Truth Tables

A truth table lists the combinations of input bits in numerical order. That is, taken together as a base 2 number, the input bits in successive rows count from 0 at the top of the table to $(2^n) - 1$ at the bottom (n being the number of input bits). In this way, the input bits themselves assign a number to each row. Since this numbering begins with 0, not 1, when we refer to row 3, we mean the fourth row in the table, the one in which the input bits comprise the base 2 representation of three (i.e., 11 in Table 1.2 or perhaps 0000011 in some larger truth tables with more input bits). Following is the truth table for AND (Table 1.3), with the base 10 representation of each row number shown.

TABLE 1.3

	x	y	$x \wedge y$
row 0 ↔	0	0	0
row 1 ↔	0	1	0
row 2 ↔	1	0	0
row 3 ↔	1	1	1

1.2.4 The Principle of Assertion

Rather than thinking of the AND function *producing* a bit, we can think of it as *being* a bit. Instead of saying x AND y is 1 only if both x and y are 1, we can say x AND y only if both x and y are equal to 1, or even x AND y only if both x and y.

Implicit in any Boolean expression is the assertion that the expression is equal to 1. If the expression is not equal to 1 (i.e., it is equal to 0) then this assertion is not true. This idea is called the principle of assertion and was first defined by the French logician Louis Couturat (1868–1914) in *Algebra of Logic* in 1905.[3]

At the time, Couturat was thinking not so much about Boolean algebra but about logic. In this field, the principle of assertion states, for example, that the statements ''All sheep have three legs'' and ''The statement 'All sheep have three legs' is true'' are equivalent statements. In the same way, the statement ''x AND y = 1'' is equivalent to the statement ''x AND y.'' The principle of assertion may seem trivial or pedantic at first, but logicians (Boole included) often draw categorical distinctions between statements about things (like sheep) and statements about statements (e.g., ''The statement '. . . .' is true''). Not only does the principle of assertion provide some greater insight into the application of Boolean algebra to the field of logic, but it will be of considerable functional use later on.

1.2.5 Some Notational Conventions

Unless otherwise noted, variables stand for bits, rather than numbers as we usually think of them. Variables are usually lowercase letters, taken from the beginning or the end of the alphabet, but may be any distinguishable symbol that is convenient. Occasionally we may create different variables by writing the same letter several times, but each time with a different number as a subscript to it. For example, x_1, x_2, x_3, and x_4 are all different variables. Such variables are particularly useful when we want to specify an arbitrarily long series of variables, for example, $x_1, x_2, x_3, \ldots, x_n$, where n is any number.

Throughout this book, we will use the notation $(x, y) = (1, 0)$ to mean x and y equal 1 and 0, respectively. So we can say that when $(x, y) = (1,1)$, $x \wedge y = 1$. This is a form of shorthand that will be used to describe specific combinations of input bits. It will be particularly useful when dealing with Boolean functions with large numbers of inputs of which we want to discuss one

[3] Couturat, a committed pacifist, was killed on the day France declared war in World War I by a truck full of French soldiers rushing to mobilize against Germany.

particular combination. For example, we can say "when $(a, b, c, d, e) = (1, 1, 0, 1, 0)$" rather than, "when $a = 1$, $b = 1$, $c = 0$, $d = 1$, and $e = 0$."

1.2.6 Circuit Symbol for AND

So far, we know that $\wedge$ means AND in algebraic expressions. The truth table shows us exactly how AND operates on its two input bits. There is one more way to represent AND that will be examined in this chapter. It is the symbol that electrical engineers use when they design digital circuits. They need to represent AND and the other Boolean functions in a visual way so that they can draw schematic diagrams (analogous to architectural blueprints) of logic circuits composed of these functions.

The devices that electronically emulate Boolean functions are microscopically tiny and are called **logic gates** rather than functions because, from an engineering perspective, the electrical signals representing 0 and 1 pass through them as they would through a gate.

In the visual language of electrical engineering, the AND gate is depicted as shown in Figure 1.1.

In a digital circuit diagram, the paths that bits travel between the gates are called **lines.** We can think of the lines as wires. The AND gate in Figure 1.1 has two input lines and one output line. The bit on the output line is 1 only if both bits on the input lines were 1.

When semiconductor circuits are built, the AND gate does not really look like this. In a computer chip, logic gates and the lines connecting them are composed of layers of different semiconducting materials and channels etched into those layers by a process known as photolithography. The circuit symbol in Figure 1.1 is the way electrical engineers represent AND gates abstractly in their logical (rather than physical) schematics. We are borrowing their visual shorthand because it shows what is going on in a given function in a graphical way when the equivalent algebraic expression might be hopelessly complicated.

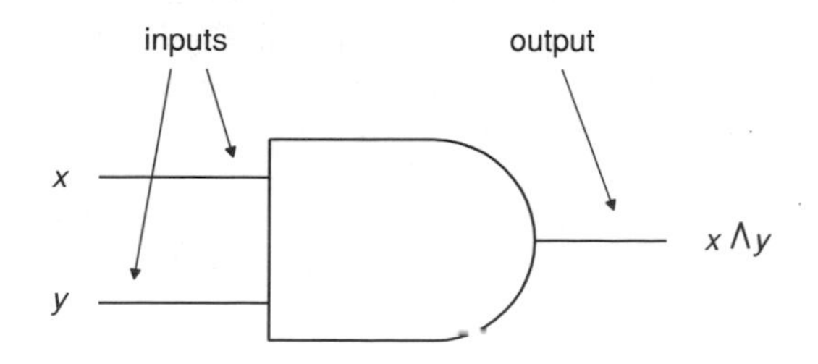

Figure 1.1 AND gate.

For example, as shown in Figure 1.2:

Figure 1.2 Examples of the operation of the AND gate.

1.3 OR

OR is the next Boolean function that we are interested in. Like AND, it takes two input bits and produces one output bit. Whereas AND asks "Are both input bits 1?'' OR asks ''Is either (or both) input bit(s) 1?''

Two bits ORed together will produce a 0 only if both are zero to start with; if either (or both) of the input bits is a 1, the result of the OR will be 1 as well. In this sense, OR is less picky than AND; in order to be 1, AND demands that both inputs are 1. If there is a 0 anywhere among its inputs, AND will be 0. OR, however, merely requires that there be at least one 1 among its two inputs in order to be 1 itself.

The algebraic symbol for OR is $\vee$. Thus

$$1 \vee 0 = 1, 0 \vee 0 = 0.$$

While ANDing a true statement with a false one yielded a false statement (''The sun is hot AND all sheep have three legs'' is a false statement), ORing a true statement and a false one is a different story: ''The sun is hot OR all sheep have three legs'' is true. Consider what happens if I flip a coin and you do not see how it lands. If I say, ''It landed heads up AND it landed tails up,'' I am lying: one of those statements has to be false, there the whole statement is false. However, if I were to say, ''It either landed heads up OR it landed tails up,'' while I would not be telling you anything you did not already know, I would be telling the truth. Only two false statements ORed together produce a falsehood, for example. ''The sun is freezing cold OR all sheep have three legs.''

In the early days of Boolean algebra, there was some argument over the OR function's handling of the case in which both inputs are true. When George Boole invented Boolean algebra, he defined the OR function in such a way that it produced a 1 if one or the other *but not both* of the inputs were 1. ORing two 1's together was not an allowed operation, much like division by zero in everyday arithmetic. Boole conceived of OR as being defined only with regard to inputs that were mutually exclusive. Boole's OR corresponds to the understanding of the word ''or'' reflected in the logical proposition, ''The Kid fled north or he fled southeast.'' While the proposition is rendered true by either of its constituent clauses being true, they could not both be true at the same time.

Later logicians, principally William Stanley Jevons (1804–1863), decided that an OR function that *includes* the case in which both inputs are 1 was more useful than Boole's original OR, which *excludes* this case. Extending and refining Boole's work, they defined OR to produce a 1 of either *or both* of its inputs are 1. Thus the OR function as we know it today is sometimes called **inclusive OR** because of its inclusion of the case in which both inputs are 1. Boole's original OR is still around, however, in the form of a function that acts like OR except that 1 OR 1 is defined to be equal to 0. This function is called **exclusive OR,** and will be discused later.

1.3.1 Truth Table for OR

To specify explicitly how the OR function operates on all possible combinations of input bits, we will write out its truth table. First enter the column headings for the two input columns and the output column and fill in the input columns with all possible combinations of two bits (Table 1.4).

TABLE 1.4

x	y	$x \vee y$
0	0	
0	1	
1	0	
1	1	

Fill in the output column on each row with the OR of the two bits in that row's input columns (Table 1.5).

TABLE 1.5

x	y	$x \vee y$
0	0	0
0	1	1
1	0	1
1	1	1

1.3.2 Circuit Symbol for OR

Like AND, OR has a symbol that electrical engineers use to represent it when they are designing digital circuits. It is called an OR gate and is shown in Figure 1.3.

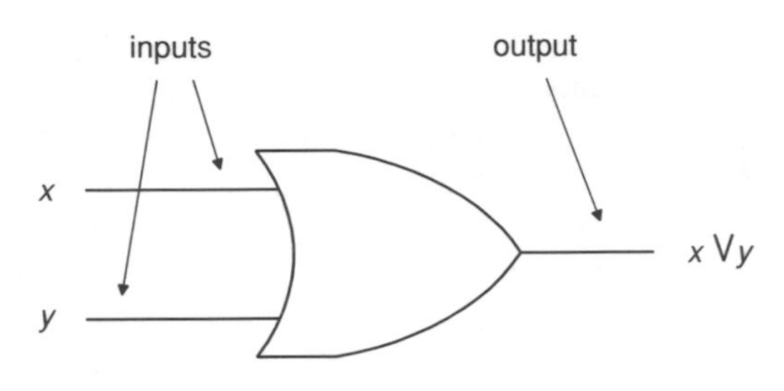

Figure 1.3 OR gate.

In the Boolean algebra literature, an analogy often is drawn between the Boolean AND and OR functions and the multiplication and addition functions in ordinary arithmetic. Some texts even use $\cdot$ and $+$ to mean AND and OR in Boolean algebraic expressions instead of $\wedge$ and $\vee$. This deliberate blurring of the line between Boolean algebra and the ordinary algebra of numbers originated with Boole himself who wanted to maintain as strong a parallel as possible between his system of logic and a system of ordinary algebra that only dealt with the integers 0 and 1.[4]

The analogy certainly seems to hold for AND and multiplication, since a multiplication table with only the integers 0 and 1 is identical to the truth table for the Boolean AND function. An addition, given the restriction that Boole imposed on the OR function, (i.e., that it should never be given two 1's as input) the addition table for the numbers 0 and 1 is the same as OR's truth table. However, throughout this book this sort of analogy for the most part will be avoided. For the purpose of learning Boolean algebra it is best not to think of the bits 0 and 1 as the integers 0 and 1 used in everyday arithmetic because attempts to draw connections between the two could lead to confusion and false assumptions. However, there are some exceptions from a notational standpoint. At times, for the sake of clarity and ease, the AND symbol ($\wedge$) will be omitted in more complex expressions. For example, "$a \wedge b$" can be written simply as "*ab*." This convention is borrowed from ordinary algebra, in which, for example, "$2x$" means "2 times x" and "xyz" means "x times y times z."

In light of the current definition of the OR function, which no longer corresponds as well as Boole's original OR function to the arithmetic addition function ($1 \vee 1 = 1$, while $1 + 1 = 2$), a more apt arithmetic analogy to Boolean AND and OR would be to view AND as the minimum function and OR as the maximum function. That is, interpreting 0 and 1 as ordinary integers, x AND y is the lesser

[4] In fact, Boole uses $\times$ and $+$ to mean AND and OR respectively in his *Investigation of the Laws of Thought.* He even attempted to develop logical parallels of subtraction and division with his system. These functions were not as useful or as well defined as AND and OR, however, and were discarded by later logicians.

of x and y, while x OR y is the greater of x and y. This interpretation accords perfectly with the truth tables for the Boolean AND and OR functions.

1.4 NOT

The last basic Boolean function to consider is NOT. It is different from AND and OR in that they are **binary functions:** they take two input bits and produce one output bit. NOT is a **unary function:** it operates on only one input bit. Essentially, NOT flips bits. It turns a 0 into a 1 and a 1 into a 0. It is also said to **invert, negate,** or **complement** bits. NOT is sometimes called the complementation function, and the opposite of a bit is called its **complement** or **inverse.** Just as we say "x AND y" we say that NOT x is the complement of x. NOT 0 is 1. Electrical engineers call a NOT gate an **inverter** because it logically inverts the electrical signal entering it.

Algebraically, NOT is represented by a short horizontal bar over the expression being complemented: $\bar{0} = 1$. The complement of x is represented by $\bar{x}$. If $x = 0$, then $\bar{x} = 1$. If $x = 1$, then $\bar{x} = 0$. Sometimes $\bar{x}$ is called "x bar" instead of "NOT x."

Just as the principle of assertion tells us that the Boolean expression x implies the assertion of its own truth ($x = 1$), it tells us that the expression $\bar{x}$ implies the assertion of x's falsehood. That is, the Boolean expression $\bar{x}$ can be thought of as a shorthand form of the assertion $x = 0$. If x is 1, then this assertion is not true, and $\bar{x}$ is equal to 0. However, if x is 0, the assertion $x = 0$ is true, and $\bar{x} = 1$.

In terms of logic, the principle of assertion tells us that the statement, "The sun is freezing cold" is equivalent to the statement, "The statement 'The sun is freezing cold' is true." Similarly, the complement of this statement is "The statement 'The sun is freezing cold' is not true" or more conveniently, "The sun is not freezing cold." We will make use of the principle of assertion, especially with regard to its implications for complemented Boolean functions, in later chapters.

1.4.1 Truth Table for NOT

NOT has a simple truth table because it only has one input bit. AND and OR each took two bits in, so their truth tables each had four rows (all possible combinations of two bits $= 2^2 = 4$). NOT's truth table only needs all possible combinations of one bit or $2^1 = 2$ rows. It also only has two columns, one for the input bit and one for the output bit. (Table 1.6).

TABLE 1.6

x	$\bar{x}$
0	1
1	0

1.4.2 Circuit Symbol for NOT

The circuit symbol for NOT is shown in Figure 1.4.

To draw a NOT gate right next to some other gate, simply draw it as a small circle (Fig. 1.5).

Consider Figure 1.6. This construction of two NOT gates connected together does nothing. Whatever bit comes into it, either a 0 or a 1, gets flipped by the first NOT gate and becomes its complement. It then gets flipped back to its original state by the second NOT gate. Thus, the bit that comes out of the second

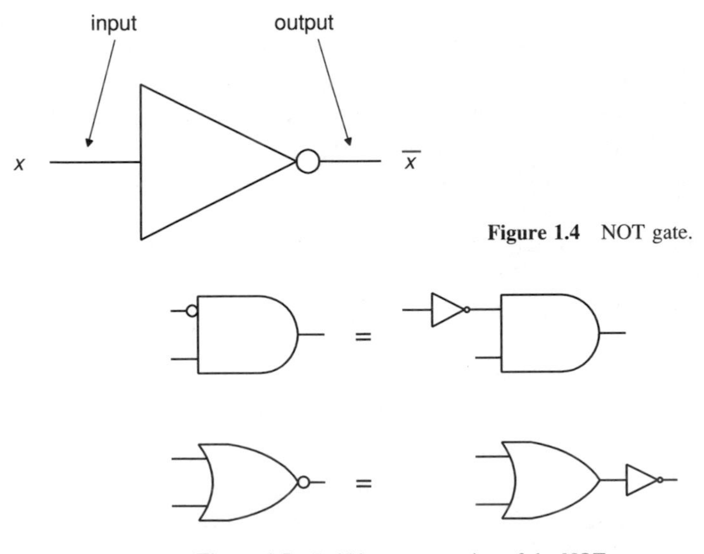

Figure 1.4 NOT gate.

Figure 1.5 Bubble representation of the NOT gate.

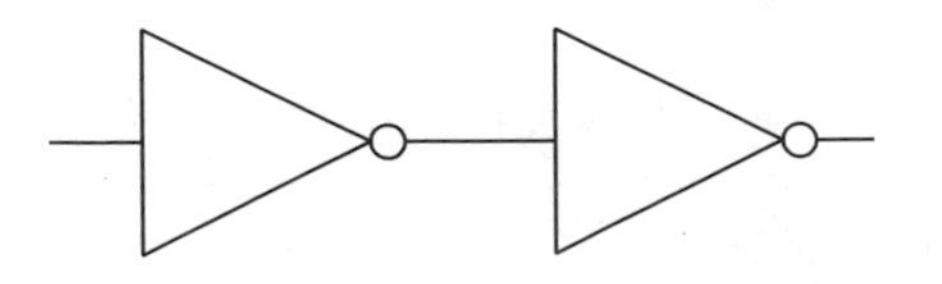

Figure 1.6 Two NOT gates connected together.

NOT gate is the same as the bit that went into the first NOT gate. If you were to string any number of NOT gates together, end to end, they would all collectively do nothing if there were an even number of them, and they would all function as one big NOT gate if there were an odd number of them. Even numbers of NOT gates connected together cancel themselves out.

EXERCISE 1.1

A. What is the value of the output bit in each of the following circuit diagrams?

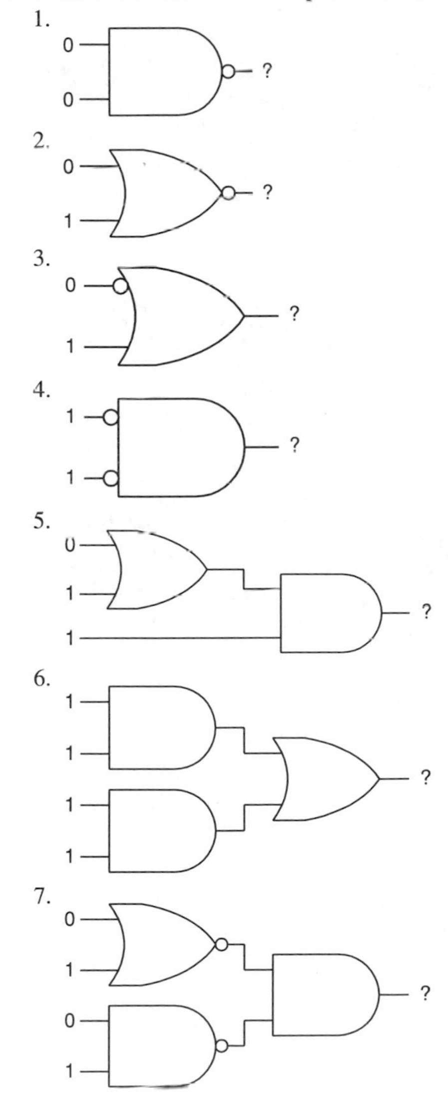

8.

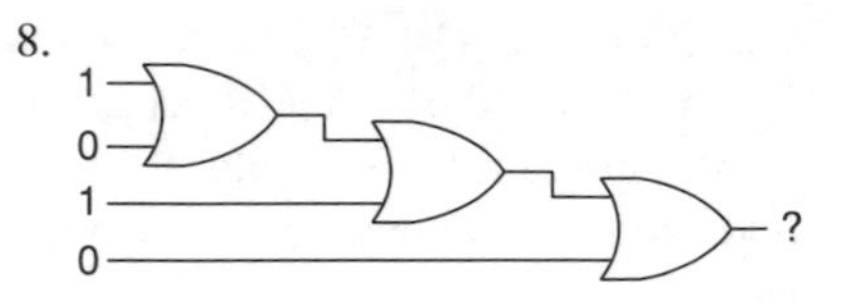

B. If possible, determine the value of the missing input bit in each of the following circuit diagrams:

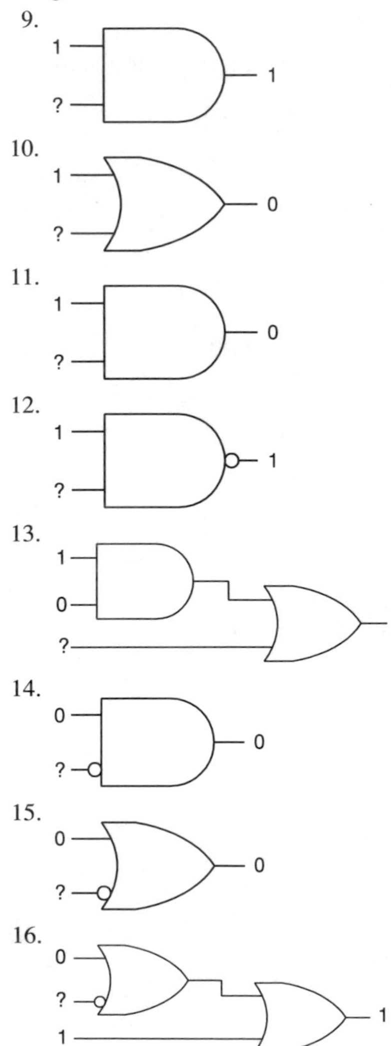

2

Combinational Logic

When a new Boolean function is created out of the old, "primitive" functions. (AND, OR, and NOT), the result is called **combinational logic** because the functions were combined to create the new function. The new function is said to be **realized** in terms of the primitive functions. Often, there are several different ways of connecting primitive logic functions to create the same new function. These different configurations of primitive functions are called different **realizations** of the new function. Especially when we show combinational logic in circuit diagram form, we call the resulting function a **combinational logic network.** Some combinational logic networks were presented in the exercises at the end chapter 1.

2.1 AND AND NOT

Consider the combinational logic network shown in Figure 2.1. How does this new Boolean function behave? It is an AND function, but one of its inputs, x, is negated before being ANDed with the other input, y. Thus, the algebraic expres-

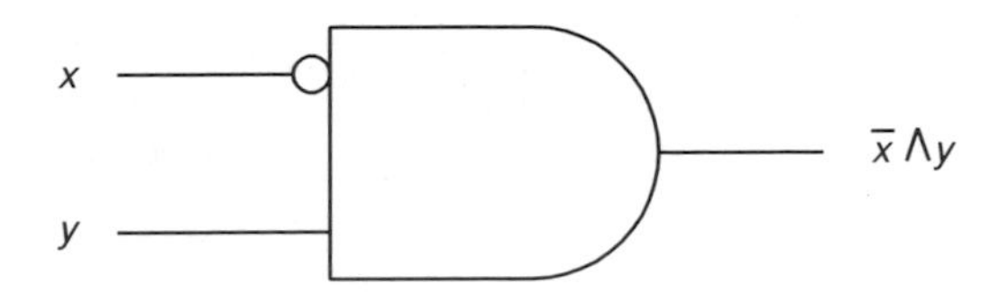

Figure 2.1 A combinational logic network with AND.

sion that describes this logic network is $\bar{x} \wedge y$; we would say, "NOT *x* AND *y*."

AND is only 1 when both of its inputs are 1. Since *x* is inverted before the AND gate in our logic network ever sees it, the only way for the AND gate to get two 1s as input is for *x* to start out as a 0, and *y* to be a 1. Then the NOT gate will switch the *x* from a 0 to a 1 and the AND gate will get two 1's as input and produce a 1 as output. Thus, this function will only be 1 when $(x, y) = (0, 1)$, and it otherwise will be 0.

When we encounter or invent a new Boolean function, it is important to write its truth table to see how the function operates on all possible inputs. Only then will we know precisely what the function does in all cases. As we write the truth table for this new function, $\bar{x} \wedge y$, we start by writing each possible combination of the two input bits, *x* and *y*, on a separate row under the headings "*x*" and "*y*" (Table 2.1).

TABLE 2.1

x	y	
0	0	
0	1	
1	0	
1	1	

We then add a column under the heading $\bar{x}$. This is to be an intermediate column, a stepping stone on the way to the output column. On each row, the entry in this column is the complement of whatever is in the *x* column on that row (note that the value of the *y* input has no effect whatsoever on the values in this intermediate column) (Table 2.2).

TABLE 2.2

x	y	$\bar{x}$
0	0	1
0	1	1
1	0	0
1	1	0

Finally, we write a column for $\bar{x} \wedge y$. In each row in this column we AND the bit in the *y* column with the bit in the $\bar{x}$ column (Table 2.3).

We wrote out the middle column, the one representing $\bar{x}$, to help us write the output column. However, once we have written the entire truth table for this

TABLE 2.3

x	y	$\bar{x}$	$\bar{x} \wedge y$
0	0	1	0
0	1	1	1
1	0	0	0
1	1	0	0

TABLE 2.4

x	y	$\bar{x} \wedge y$
0	0	0
0	1	1
1	0	0
1	1	0

function, all we care about is the bit that appears in the output column for each combination of the two input bits, and we can eliminate the middle column. Table 2.4 is the truth table for our new function.

2.2 GROUPING WITH PARENTHESES

Now consider the combinational logic network in Figure 2.2.

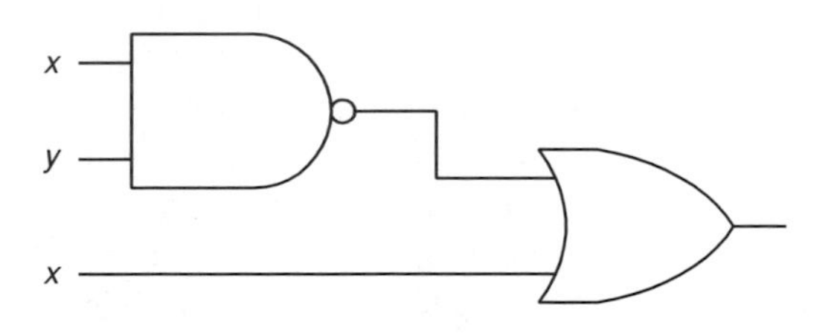

Figure 2.2 A combinational logic network with two x.

While this circuit diagram appears to have three inputs, two of them are labeled x and the other is labeled y. This means that whatever bit goes into one x input must also go into the other x input. They are always the same. For this reason, the circuit diagram really has only two inputs. A clearer illustration of this is shown in Figure 2.3.

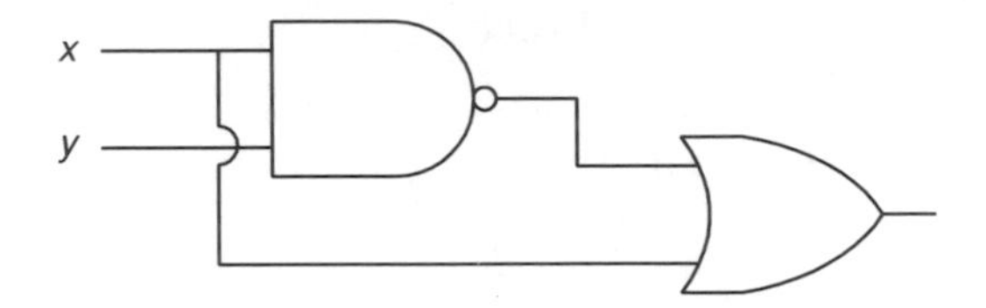

Figure 2.3 A combinational logic network with *a*.

Both ways of drawing the circuit diagram are equally correct; the second way just shows more graphically that the circuit described really has only two inputs, not three. The little "jump" in the x line as it crosses the y line is something that electrical engineers use to indicate that while one wire crosses over another, they do not actually touch electrically; bits cannot flow from one of the lines onto the other.

How is the algebraic expression that represents this logic network written? In particular, how do we indicate that we want to AND x and y together first, then invert the output bit, and then OR it with x?

As in ordinary algebra, we can group quantities together with parentheses. For example, according to the laws of ordinary arithmetic, if we had to evaluate the expression $(4 + 2) \times 5$, we would solve the part in parentheses first:

$$(4 + 2) \times 5 =$$

$$6 \times 5 =$$

$$30.$$

In similar fashion, the algebraic expression corresponding to the circuit diagram above is $\overline{(x \wedge y)} \vee x$. The NOT bar over the parenthetical expression $(x \wedge y)$ means that we evaluate $(x \wedge y)$ first, then NOT the resulting output bit. Finally, we feed the output bit from $\overline{(x \wedge y)}$ into an OR gate along with the second x bit. The output bit from the OR is the output bit from our entire function.

Now we will write the truth table for this new function, $\overline{(x \wedge y)} \vee x$. As always, we start the truth table by writing the headings of the input columns, x and y. Below these we list all possible combinations of inputs, one combination per row. Because there are two binary inputs and there are $2^2 = 4$ possible combinations of two bits, there are four rows in this truth table (Table 2.5).

We then create a new column to represent the part of the expression that is in parentheses, $x \wedge y$. In each row, we fill in this column with the result of ANDing together the x and y bits in that row Table 2.6.

So far, this is our standard AND truth table. Next, we create a column to represent $\overline{(x \wedge y)}$. This column contains the complement of each bit in the $(x \wedge y)$ column Table 2.7.

Finally, we create a column for the entire expression, $\overline{(x \wedge y)} \vee x$. In this

TABLE 2.5

x	y	
0	0	
0	1	
1	0	
1	1	

TABLE 2.6

x	y	$x \wedge y$
0	0	0
0	1	0
1	0	0
1	1	1

TABLE 2.7

x	y	$x \wedge y$	$\overline{(x \wedge y)}$
0	0	0	1
0	1	0	1
1	0	0	1
1	1	1	0

column, we OR together the bits in the x column on the far left of the table and the bits in the $\overline{(x \wedge y)}$ column (Table 2.8).

The final truth table for the expression $\overline{(x \wedge y)} \vee x$, without the clutter of the intermediate columns is shown in Table 2.9.

This is a strange function. There is a 1 in the output column in every row

TABLE 2.8

x	y	$x \wedge y$	$\overline{(x \wedge y)}$	$\overline{(x \wedge y)} \vee x$
0	0	0	1	1
0	1	0	1	1
1	0	0	1	1
1	1	1	0	1

TABLE 2.9

x	y	$\overline{(x \wedge y)} \vee x$
0	0	1
0	1	1
1	0	1
1	1	1

of the truth table. This means that no matter what bits this function receives as inputs, it will always produce a 1 as output. Effectively, it ignores its inputs completely, and for this reason it is not a particularly interesting function. A Boolean function that produces a 1 as its output bit for all possible combinations of input bits is called a **tautology.** "Tautology" is a perfectly good conversational English word as well. It is usually used dismissively to describe a statement or argument that follows obviously from its own premises and thus is redundant.

How about the function $(x \vee y) \wedge (\bar{y} \vee z)$? First, notice that it has two expressions in parentheses that are each evaluated separately and then combined by the AND in the middle. Second, notice that the variable y appears twice in this expression (although the second time it appears it is inverted). Thus, there are really only three inputs, not four: x, y, and z.

Before we write out the truth table for this expression, we will construct the circuit diagram, but we will do it in stages. Figure 2.4 shows the diagram for $x \vee y$.

Now for $\bar{y} \vee z$ (Fig. 2.5) we AND the two parts together as shown in Figure 2.6, or alternatively as shown in Figure 2.7.

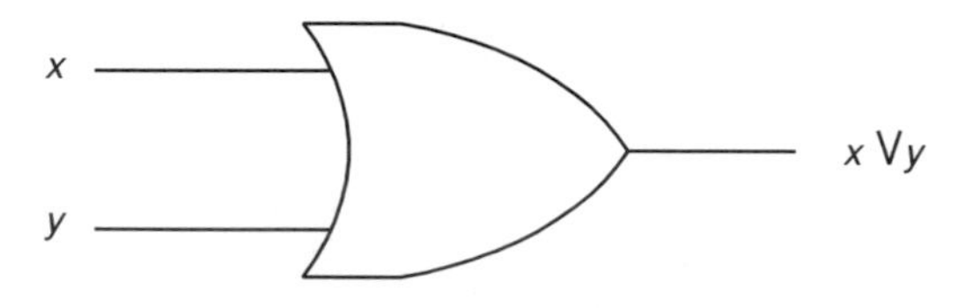

Figure 2.4 $x \vee y$.

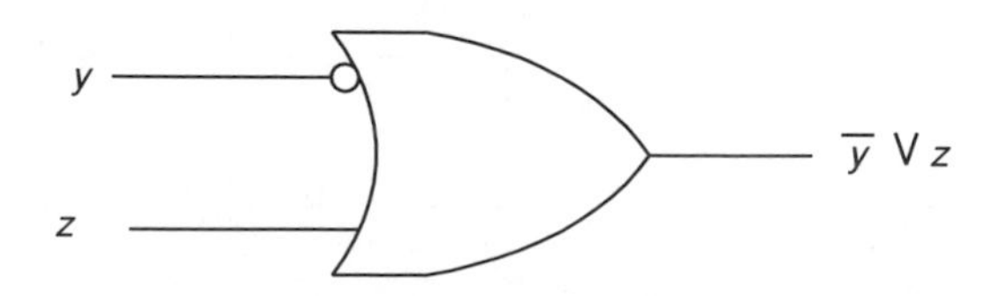

Figure 2.5 $\bar{y} \vee z$.

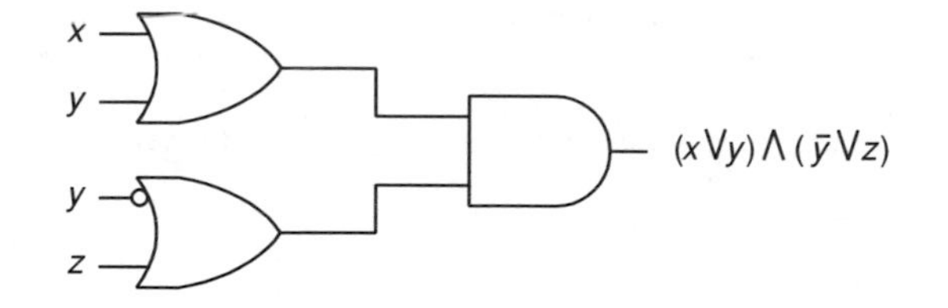

Figure 2.6 $(x \vee y) \wedge (y \vee z)$.

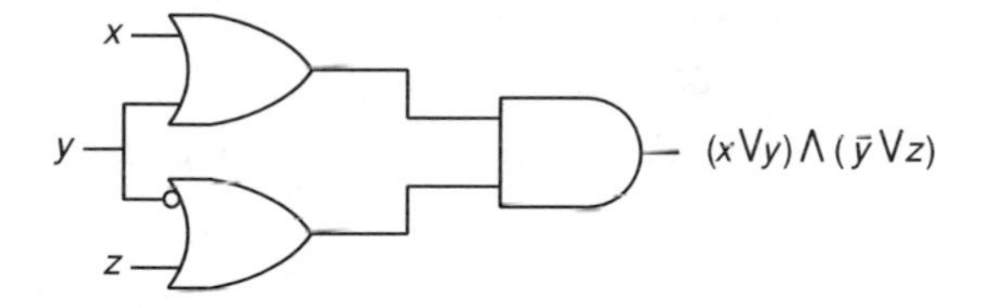

Figure 2.7 $(x \vee y) \vee (y \vee z)$ with split y input.

Now we will write out the truth table as shown in Table 2.10. There are three variables, so we need three input bit column headings, x, y, and z, and one row for each of the $2^3 = 8$ combinations of three bits.

TABLE 2.10

x	y	z	
0	0	0	
0	0	1	
0	1	0	
0	1	1	
1	0	0	
1	0	1	
1	1	0	
1	1	1	

We need to create separate columns for $x \vee y$, $\bar{y}$, and $\bar{y} \vee z$ in Table 2.11.

While this is a three-input function, each of these individual subfunctions contains fewer than three input bits. To write the output columns for these subfunctions, we ignore those of the function's input bit(s) that are not needed to determine the value of a particular subfunction. For example, $x \vee y$ has a 0 in

TABLE 2.11

x	y	$\bar{z}$	$x \vee y$	$\bar{y}$	$\bar{y} \vee z$
0	0	0	0	1	1
0	0	1	0	1	1
0	1	0	1	0	0
0	1	1	1	0	1
1	0	0	1	1	1
1	0	1	1	1	1
1	1	0	1	0	0
1	1	1	1	0	1

its output column in all rows in which $(x, y) = (0, 0)$, of which there are two, $(x, y, z) = (0, 0, 0)$ and $(x, y, z) = (0, 0, 1)$. For the purposes of determining the value of $x \vee y$ for a particular row in the truth table, we do not care what z is. By the same token, the column for $\bar{y}$ simply contains the inverse of y in each row, with no regard for the values of x and z. This is illustrated by the circuit diagram in Figure 2.6 in which not all input lines are used as inputs to every logic gate.

Finally, we write the column for the complete expression, $(x \vee y) \wedge (\bar{y} \vee z)$, by ANDing the bits in the $x \vee y$ column with the bits in the $\bar{y} \vee z$ column as shown in Table 2.12.

TABLE 2.12

x y z	$x \vee y$	$\bar{y}$	$\bar{y} \vee z$	$(x \vee y) \wedge (\bar{y} \vee z)$
0 0 0	0	1	1	0
0 0 1	0	1	1	0
0 1 0	1	0	0	0
0 1 1	1	0	1	1
1 0 0	1	1	1	1
1 0 1	1	1	1	1
1 1 0	1	0	0	0
1 1 1	1	0	1	1

EXERCISE 2.1

A. For each of the following Boolean algebraic expressions: (a) draw the corresponding circuit diagram and (b) write the truth table. Use extra columns in the truth tables to keep track of intermediate results.

1. $\bar{x} \vee y$
2. $(x \wedge \bar{y}) \wedge \bar{z}$
3. $(x \wedge \bar{y}) \wedge \bar{x}$
4. $x \wedge \bar{x}$
5. $\bar{x} \vee [y \wedge (x \vee \bar{z})]$
6. $\overline{(x \wedge \bar{y})} \wedge (y \vee \bar{z})$
7. $(x \wedge y) \vee (x \wedge \bar{y})$
8. $\bar{x} \wedge y$
9. $[(\bar{a} \wedge b) \vee c] \wedge (c \vee \bar{d})$
10. $[(x \vee \bar{y}) \wedge (x \vee z)] \vee [\bar{x} \vee (y \wedge z)]$ (Interpret your result.)
11. $\overline{(x \wedge y)} \vee [(x \vee \bar{y}) \wedge (\bar{x} \vee y)]$

B. For each of the following circuit diagrams (a) write the corresponding Boolean algebraic expression and (b) the truth table. Use extra columns in the truth tables to keep track of intermediate results.

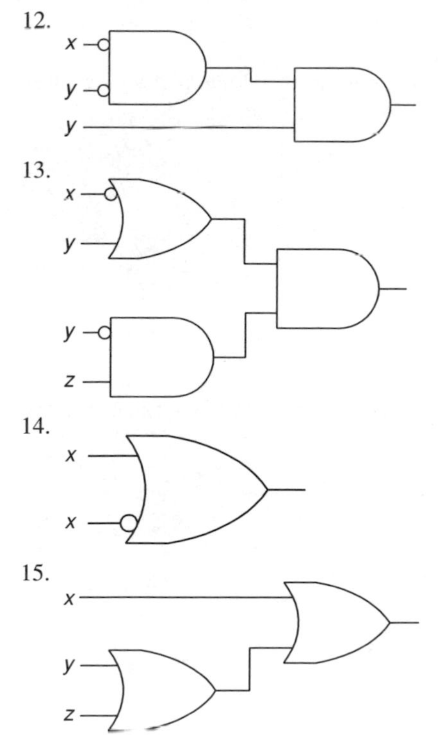

16.

17.

18.

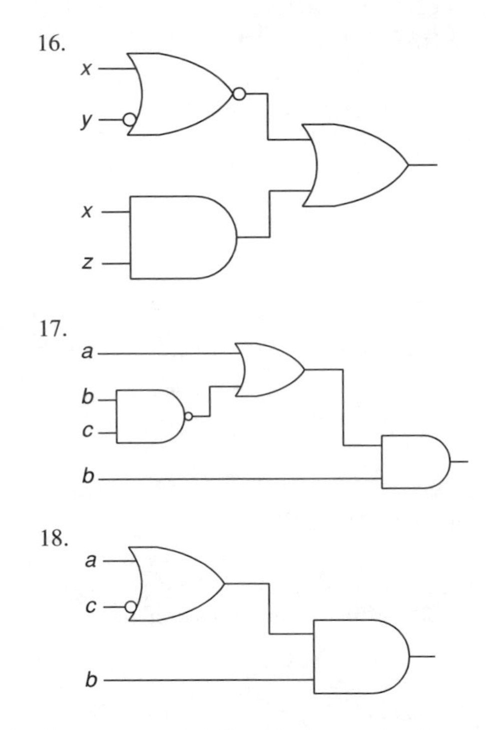

C. Compare your result with that obtained in exercise 17 and interpret your comparison. May we infer that $\overline{(b \wedge c)}$ equals $\bar{c}$? Clearly we may not.

19.

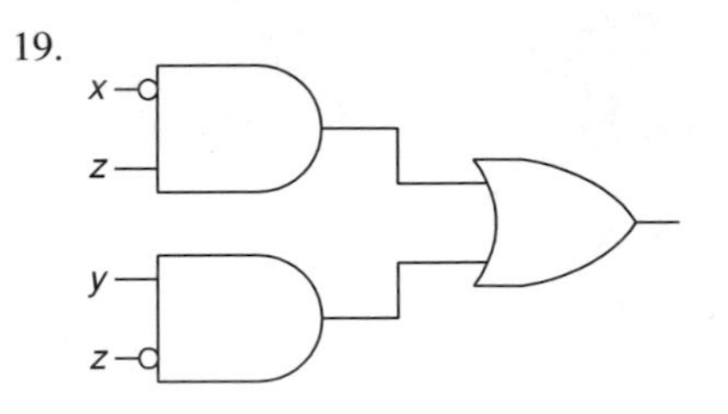

D. Write the algebraic expressions of the functions whose truth tables are presented in 20, 21, and 22.

20.

x	y	
0	0	0
0	1	1
1	0	0
1	1	0

21.

x	y	
0	0	0
0	1	0
1	0	1
1	1	1

22.

x	y	
0	0	1
0	1	0
1	0	0
1	1	1

2.3 AND AND OR WITH MORE THAN TWO INPUTS

AND and OR, as we know them, take two input bits and produce one bit as output. However, AND and OR would be more useful to us if they were extended to deal with any number of input bits. We will redefine AND and OR, then, so that they take two or *more* bits as input.

Let the output of AND be 1 if and only if *all* of its inputs are 1, and 0 otherwise; that is, if *any* of its inputs is 0, the output of AND is to be 0.

Let the output of OR be 0 if and only if *all* of its input bits are 0, and 1 otherwise; that is, if *any* of its inputs is 1, the output of OR is 1.

This extension of the definitions of AND and OR flows naturally from our intuitive understanding of them. The old two-input AND and OR may be thought of as special cases of the new arbitrary-input AND and OR. Note that if the word "both" is substituted for "all," and the words "either or both" for "any" in the redefinitions, we have the definitions of the old two-input AND and OR functions.

These new definitions also correspond to our English language understanding of the words "and" and "or." For example, the statement, "I like chocolate AND I like vanilla, AND I like fudge swirl" is true only if all three of the component statements are true. Similarly, the statement, "This rock contains cobalt OR magnesium OR uranium OR copper" is true if the rock contains any (or any combination) of the four elements. The only case in which the statement is false is that in which the rock in question contains none of the four elements listed.

We will continue to use the same circuit symbols for the arbitrary-input AND and OR gates, but now there can be more than two input lines. There will still be only one output line, as the new AND and OR still only produce one output bit.

2.4 Algebraic Examples of Arbitrary-Input AND and OR Functions

When writing algebraic expressions using the new arbitrary-input AND and OR functions, we still use the old algebraic symbols $\wedge$ and $\vee$, and we put them between all bits that are to be ANDed or ORed together. For example:

$$1 \wedge 1 \wedge 1 \wedge 1 \wedge 1 = 1.$$

Any number of 1's ANDed together yield a 1 as output. However,

$$1 \wedge 1 \wedge 0 \wedge 1 \wedge 0 \wedge 0 \wedge 1 = 0.$$

There are 0's ANDed together with the 1's in the above expression. Thus the output bit is 0. Even one 0 among many 1's is enough to cause the AND function to yield a 0 as output:

$$1 \wedge 1 \wedge 1 \wedge 1 \wedge 1 \wedge 0 \wedge 1 \wedge 1 \wedge 1 \wedge 1 = 0.$$

Similarly for OR:

$$0 \vee 0 \vee 0 \vee 0 = 0, \text{ but}$$

$$0 \vee 0 \vee 1 \vee 0 \vee 0 \vee 0 \vee 0 = 1.$$

Even a single 1 ORed together with any number of 0's (or other 1's) is enough to make the OR function yield a 1 as output.

2.5 TRUTH TABLES FOR ARBITRARY-INPUT AND AND OR FUNCTIONS

We may write truth tables for our new functions to see exactly how they operate. As an example, we will write the truth table for a four-input AND function and call our inputs *a, b, c,* and *d.* We start the truth table with the four input columns beneath their headings. Since there are four inputs, there are $2^4 = 16$ possible combinations of those inputs and thus 16 rows in the truth table (Table 2.13).

As expected (given our understanding of the AND function), there is a 0 in the output column in every row except the last one because the last row is the only row in which all four inputs are 1.

Now we will write the truth table for OR with three inputs and call them *x, y,* and *z* (Table 2.14).

Only in the first row, in which all three inputs are 0, is the output column 0. In all other rows there is at least one 1 among the input bits, which is enough to make the output from the OR function 1.

TABLE 2.13

a	b	c	d	$a \wedge b \wedge c \wedge d$
0	0	0	0	0
0	0	0	1	0
0	0	1	0	0
0	0	1	1	0
0	1	0	0	0
0	1	0	1	0
0	1	1	0	0
0	1	1	1	0
1	0	0	0	0
1	0	0	1	0
1	0	1	0	0
1	0	1	1	0
1	1	0	0	0
1	1	0	1	0
1	1	1	0	0
1	1	1	1	1

TABLE 2.14

x	y	z	$x \vee y \vee z$
0	0	0	0
0	0	1	1
0	1	0	1
0	1	1	1
1	0	0	1
1	0	1	1
1	1	0	1
1	1	1	1

2.6 CREATING ARBITRARY-INPUT AND AND OR GATES FROM THE OLD TWO-INPUT KIND

We redefined AND and OR with just a wave of the hand, declaring the old definitions for AND and OR inadequate and rewriting them to suit our purposes. This may seem like cheating, mathematically speaking (not to mention going back on the promise that you would never know anything more about the AND and OR functions than was presented in chapter 1!). It turns out that there is a more formal way to make the transition to these new definitions without shifting our entire mathematical system onto a new foundation—we can make our new AND and OR out of the old two-input AND and OR. Doing so would help to preserve the elegance of the system.

Computer programmers use the term **elegant** to describe a beautiful program—one that is clever and compact and reflects a deep understanding of the programming language and the problem being solved by the program. Elegant programs usually exhibit creative leaps of logic and a certain economy of design. However, elegance is a subtle concept, and any 20 programmers will likely have 20 different ideas as to what exactly elegance means. Similarly, mathematicians speak of elegant proofs or elegant mathematical systems. While defining an elegant mathematical system is a bit subjective, a system usually is more elegant the more complex the objects are that can be built within it, and the simpler (and fewer) the building blocks with which they are built. Mathematical systems that are not elegant might be described as clunky, awkward, or unwieldy.

From the point of view of mathematical elegance, we would like to keep our initial system of Boolean algebra as simple as possible, consisting only of the things 0 and 1, and the functions two-input AND, two-input OR, and one-input NOT. We would like to do this while sacrificing as little as possible of our power to build complex functions and prove elaborate truths within the system. If we find that we weaken the system considerably by restricting it to the two-input functions, we are confronted with a trade-off. In order to strengthen the system, we may redefine the primitive functions in the system to accommodate any number of inputs.

However, we may decide to leave the functions the way they are to streamline the system. In this case we may find that otherwise provable truths and constructable functions slip through our fingers, and we must resign ourselves to their being forever beyond the reach of our mathematical system. There is no right answer. A large portion of the art of mathematics lies in the ability to make such decisions and shape mathematical systems according to one's own sense of mathematical elegance.

In this case, however, we are given an easy way out, because it can be shown that it is possible to build the new AND and OR out of the old two-input kind. Thus, we may preserve the leanness of the mathematical system by keeping the two-input AND and OR as the primitive functions, yet we will have access to any constructions that can be built out of arbitrary-input AND and OR. In effect, we will have proven that anything we build in the future using the arbitrary-input AND and OR we could have also built with the old two-input models.

2.7 AN ARBITRARY-INPUT AND GATE

Recalling that a 1, when output from a two-input AND gate, means that neither of the AND gate's inputs is 0, consider the circuit diagram in Figure 2.8.

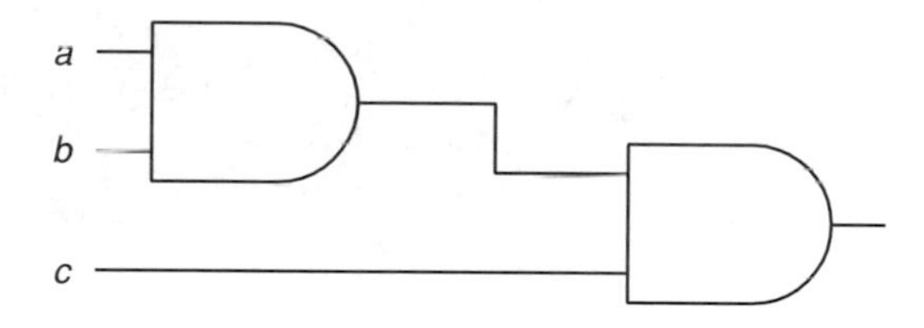

Figure 2.8 Three-input AND made of two-input ANDs.

The only way this circuit could produce a 1 is if all three input lines were 1. This is because the output bit could only be 1 if both inputs to the second AND gate are 1, which includes input c, and the only way the other input to the second AND gate could be 1 is if both inputs to the first AND gate (a and b) were both 1 as well.

The algebraic expression of this realization of the three-input AND gate is $(a \wedge b) \wedge c$. Table 2.15 is the truth table for the expression to prove that this function conforms to our definition of an arbitrary-input AND gate.

There are 0's in the output columns for all rows except the last row in which all three input bits are 1, so we have indeed created a true three-input AND gate.

TABLE 2.15

a	b	c	$a \wedge b$	$(a \wedge b) \wedge c$
0	0	0	0	0
0	0	1	0	0
0	1	0	0	0
0	1	1	0	0
1	0	0	0	0
1	0	1	0	0
1	1	0	1	0
1	1	1	1	1

Just as we expanded the two-input AND gate to three inputs by ANDing its output with the third input, we can create a four-input AND gate by feeding the first three inputs into a three-input AND gate, then ANDing the output of that with the fourth input bit as shown in Figure 2.9.

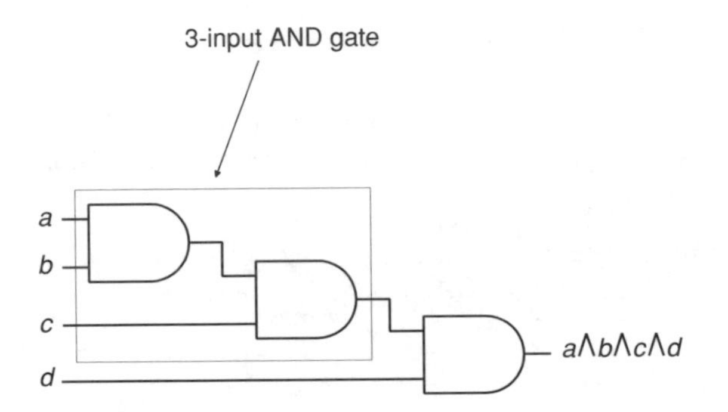

Figure 2.9 Four-input AND made of two-input ANDs.

A 1 from the three-input AND gate means that none of the three inputs is 0, and only if that is true and the fourth input is also a 1 will the final output be a 1. The algebraic expression for this (in terms of two-innut ANDs) is $[(a \wedge b) \wedge c] \wedge d$.

We could go on forever, building AND gates with successively more input bits by ANDing each additional input bit with the output bit from the previously built AND gate. So for any number n greater than two, we can make an n-input AND gate by ANDing the nth input bit with an $(n\text{-}1)$-input AND gate. This is illustrated by the circuit diagram shown in Figure 2.10.

To build an $(n\text{-}1)$-input AND gate we go around again, ANDing the $(n\text{-}1)$th input with the output from an $(n\text{-}2)$-input AND gate, until we end up ANDing the third input with the output from the familiar primitive 2-input AND gate.

Here, then, is the algebraic definition of the arbitrary AND function in terms of 2-input ANDs:

$$x_1 \wedge x_2 \wedge x_3 \wedge x_4 \ldots x_n = (\ldots(x_1 \wedge x_2) \wedge x_3) \ldots \wedge x_n).$$

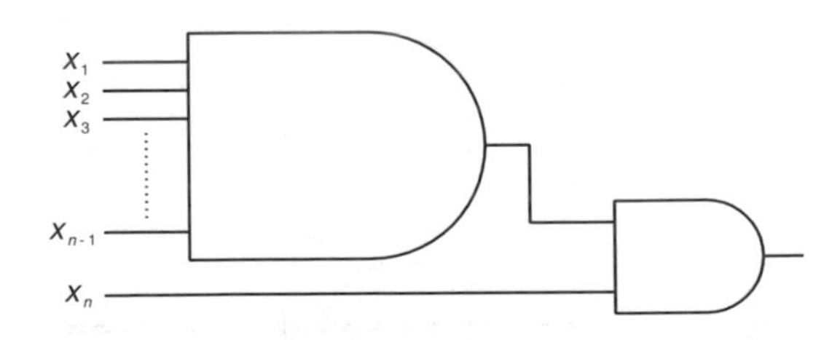

Figure 2.10 n-input AND gate where n is any number.

Whenever we build new functions out of arbitrarily long chains of other functions we say we have **cascaded** the smaller functions together. This is because the arrangement is like a waterfall with steps in it that the water hits and bounces off with bits flowing from one gate to the next on down the chain. In this case we have cascaded the two-input AND gates together to make AND gates with any desired number of inputs.

2.8 AN ARBITRARY-INPUT OR GATE

We can cascade two-input ORs together to make arbitrary-input ORs in the same way (Fig. 2.11).

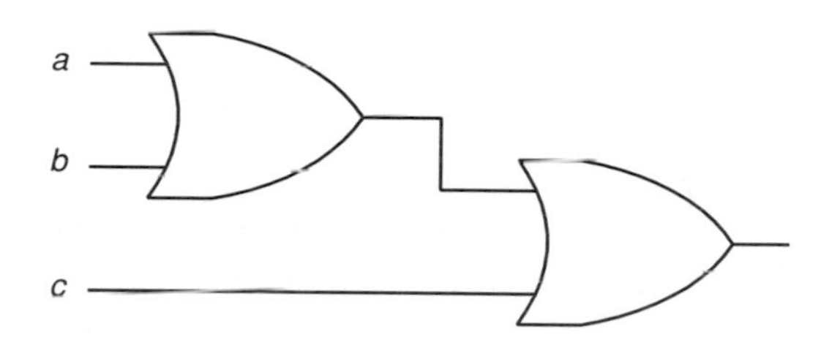

Figure 2.11 Three-input OR made of two-input ORs.

This combinational logic network is represented algebraically by the expression $(a \bigvee b) \bigvee c$. Its output bit will be 1 as long as at least one of the two inputs to the second OR gate is 1, that is, if either input c or the output bit of the first OR gate, or both, are 1. The output of the first OR gate will be 1 as long as at least one of inputs a and b is 1. The only way the final output could be 0 is if all three inputs, a, b, and c, are all 0 as well. This satisfies our definition of an arbitrary-input OR gate. However, to prove it we will write out the truth table as shown in Table 2.16.

There are 1's in the output column for all rows except the first one in which all three input bits are 0. Thus, this is a true three-input OR function.

As with the AND function, we can create an OR function with as many inputs as we want by cascading two-input OR functions together. As with AND, if we have an OR function with n-1 inputs, we can create an OR function with n inputs by ORing the output of the n-1 input OR function with the nth input bit (for any number n greater than two) (Fig. 2.12). Starting with two-input ORs we can make a three-input OR function (as we just did above). By ORing its output with a fourth input bit we can create a four-input OR function, and so on. We will define an arbitrary-input OR gate and function as follows:

$$x_1 \bigvee x_2 \bigvee x_3 \bigvee \ldots \bigvee x_n = (\ldots(x_1 \bigvee x_2) \bigvee x_3) \bigvee \ldots \bigvee x_n).$$

TABLE 2.16

a	b	c	$a \vee b$	$(a \vee b) \vee c$
0	0	0	0	0
0	0	1	0	1
0	1	0	1	1
0	1	1	1	1
1	0	0	1	1
1	0	1	1	1
1	1	0	1	1
1	1	1	1	1

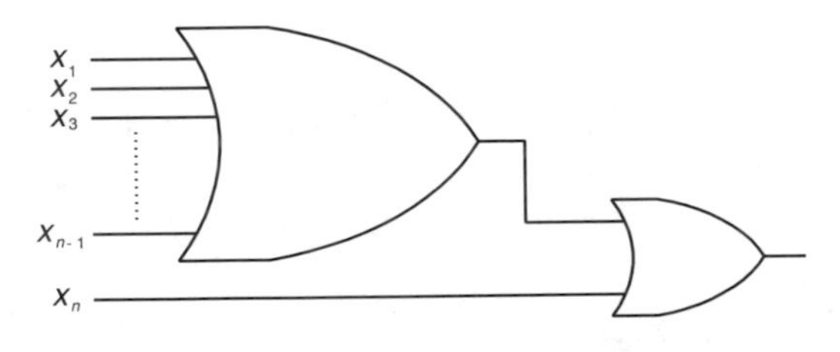

Figure 2.12 n-input OR gate where n is any number.

Throughout the rest of this book, unless stated otherwise, AND and OR are assumed to mean AND and OR with an arbitrary number of inputs.

As an illustration, consider the Boolean function $(x \vee y \vee z) \wedge (x \vee y \vee \bar{z}) \wedge \overline{(x \vee \bar{y})}$. It has three inputs, x, y, and z, and uses three-input functions in several places. Its circuit diagram is shown in Figure 2.13.

To write the truth table for this function, we must set up the standard input columns and headings for a three-input truth table as shown in Table 2.17.

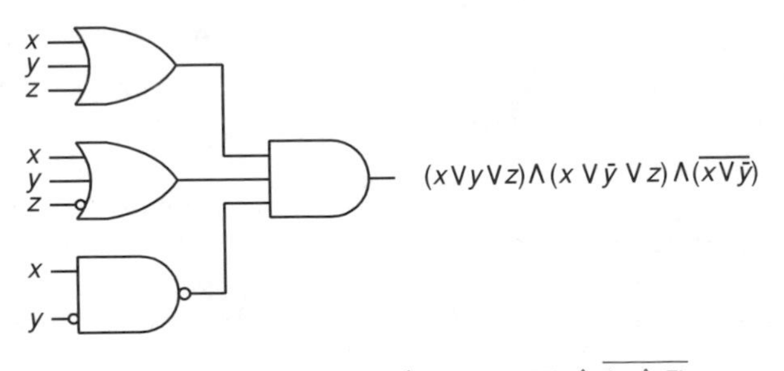

Figure 2.13 $(x \vee y \vee z) \wedge (x \vee y \vee \bar{z}) \wedge \overline{(x \wedge \bar{y})}$.

TABLE 2.17

x	y	z	
0	0	0	
0	0	1	
0	1	0	
0	1	1	
1	0	0	
1	0	1	
1	1	0	
1	1	1	

Now, as an intermediate step, we create the additional columns in Table 2.18.

TABLE 2.18

x	y	z	$\bar{y}$	$\bar{z}$	$x \wedge \bar{y}$	$\overline{(x \wedge \bar{y})}$
0	0	0	1	1	0	1
0	0	1	1	0	0	1
0	1	0	0	1	0	1
0	1	1	0	0	0	1
1	0	0	1	1	1	0
1	0	1	1	0	1	0
1	1	0	0	1	0	1
1	1	1	0	0	0	1

So far, this is more or less similar to the examples we dealt with in the previous section. The new wrinkle appears with the columns for $(x \vee y \vee z)$ and $(x \wedge y \vee \bar{z})$. In these columns, we put the results of a three-input OR of the bits involved as shown in Table 2.19.

Finally, we perform a three-input AND on the $\overline{(x \wedge y)}$, $(x \vee \bar{y} \vee z)$, and $(x \vee y \vee \bar{z})$ columns as shown in Table 2.20. Without the intermediate columns, the truth table looks like Table 2.21.

Look closely at the output column. In each row, the output bit is equal to the bit in the y input column. This is another deceptively uninteresting function No matter what inputs it receives, its output is always the same as the y input. It essentially ignores x and z and passes y through untouched. It may strike you

TABLE 2.19

x y z	$\bar{y}$	$\bar{z}$	$x \wedge \bar{y}$	$\overline{(x \wedge \bar{y})}$	$x \vee y \vee z$	$x \vee y \vee \bar{z}$
0 0 0	1	1	0	1	0	1
0 0 1	1	0	0	1	1	0
0 1 0	0	1	0	1	1	1
0 1 1	0	0	0	1	1	1
1 0 0	1	1	1	0	1	1
1 0 1	1	0	1	0	1	1
1 1 0	0	1	0	1	1	1
1 1 1	0	0	0	1	1	1

TABLE 2.20

$\overline{(x \wedge \bar{y})}$	$x \vee y \vee z$	$x \vee y \vee \bar{z}$	$\overline{(x \wedge \bar{y})} \wedge (x \vee y \vee z) \wedge (x \vee y \vee \bar{z})$
1	0	1	0
1	1	0	0
1	1	1	1
1	1	1	1
0	1	1	0
0	1	1	0
1	1	1	1
1	1	1	1

TABLE 2.21

x	y	z	$\overline{((x \wedge \bar{y}))} \wedge (x \vee y \vee z) \wedge (x \vee y \vee \bar{z})$
0	0	0	0
0	0	1	0
0	1	0	1
0	1	1	1
1	0	0	0
1	0	1	0
1	1	0	1
1	1	1	1

that the function is needlessly complex for something that only passes one of its inputs through. If so, you are correct, and we will soon learn ways to reduce such a function to its bare essentials, stripping away the unnecessary complications without resorting the scrutiny of the truth table as we did in this case.

EXERCISE 2.2

A. For the each of the following Boolean algebraic expressions: (a) draw the corresponding circuit diagram and (b) write the truth table. Use extra columns in the truth tables to keep track of intermediate results.

1. $a \vee b \vee \bar{c} \vee d$
2. $(a \wedge \bar{b} \wedge \bar{c}) \vee (b \wedge c) \vee (\bar{a} \wedge c)$
3. $(\bar{x} \wedge y) \vee (x \wedge \bar{y}) \vee (x \wedge y)$
4. $\overline{(x \wedge \bar{y} \wedge z)} \wedge (\bar{x} \wedge z)$
5. $\overline{(a \vee \bar{b} \vee c)}$
6. $(a \vee \bar{b} \vee c) \wedge \bar{a} \wedge b$

B. Compare the outputs of the functions given in 5 and 6. Interpret the result.

C. For each of the following circuit diagrams write (a) the corresponding Boolean algebraic expression and (b) the truth table. Use extra columns in your truth tables to keep track of intermediate results.

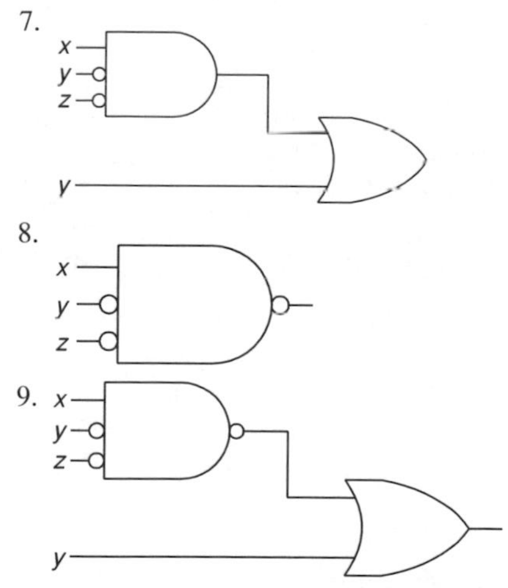

D. Compare the outputs of the functions given in the preceding exercises.

E. Interpret the result.

10.

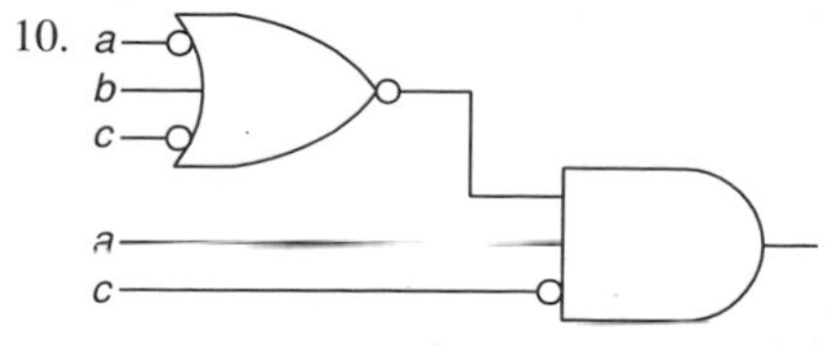

11.

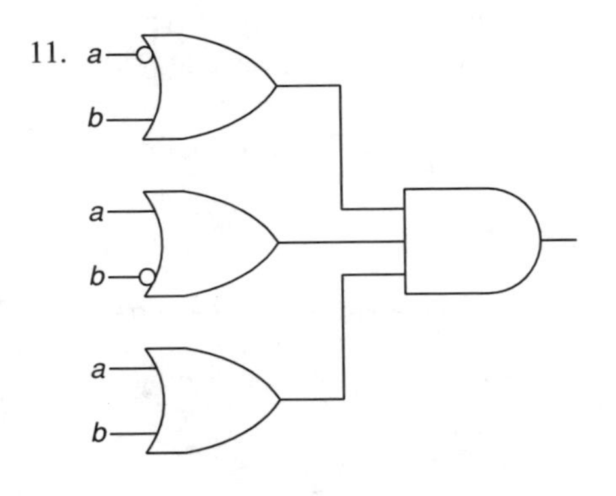

3

The Algebra of Sets and Venn Diagrams

In this chapter we will explore a mathematical system called the **algebra of sets.** Its relevance to Boolean algebra will become apparent as we proceed. As with all mathematical systems, we must begin by defining some things that the system deals with and some functions that operate on those things. Then we prove various properties of our things and the functions that operate on them. For the algebra of sets, the fundamental thing we deal with is called a **set.**

3.1 THE SET

Essentially, a set is a grouping of objects. Any object you can think of may be inside a given set or outside of it. For example, consider the set consisting of all objects that are made of wood. We cannot enumerate all of the countless wooden objects, but we can say that my desk (which is oak) is inside this set. We can also say that the tires on my car are outside of it. However, if we turn our consideration to the set of all things that are round, we notice that my tires are in this set, while my desk is not. Both items are inside the set of things that belong to me. A thing that is inside a set is called an **element** of that set.

The concept of the set is fairly abstract. A set is not a collection of things in the sense that the things within the set are physically in one place or organized in any particular way. The elements of a set need not be tangible, and there can be an infinite number of them, as in the case of the set of odd integers or the set of all fractions (rational numbers) that exist between the numbers 23 and 24.[1] The branch of mathematics called set theory has been rocked by complications arising from the implications of sets of sets.[2] For our purposes, however, the important characteristic of a set is the ability to determine whether any given object is inside or outside of the set, which is to say, whether or not the object is an element of the set.

3.2 VENN DIAGRAMS

To a great extent, George Boole conceived of his Boolean algebra as an algebra of sets. While many of the examples in his work deal with logical propositions and their truth or falsehood, he also devoted a good deal of effort to explaining his ideas in terms of objects being included or excluded from sets. This idea was developed further by another Englishman, Reverend John Venn (1834–1923). Although Venn's conception of sets (or classes, as they were sometimes known) and the rules of their manipulation differed only incrementally from Boole's, he did develop a novel way of representing them abstractly using diagrams. Venn diagrams, as they have been known ever since, remain John Venn's most lasting contribution to the field of logic.

A Venn diagram begins with a box that represents the **Universe,** in the sense that in the shorthand notation of Venn diagrams, all possible objects in the universe lie within this box. The Universe is the set that contains everything and has its own symbol, U (see Fig. 3.1).

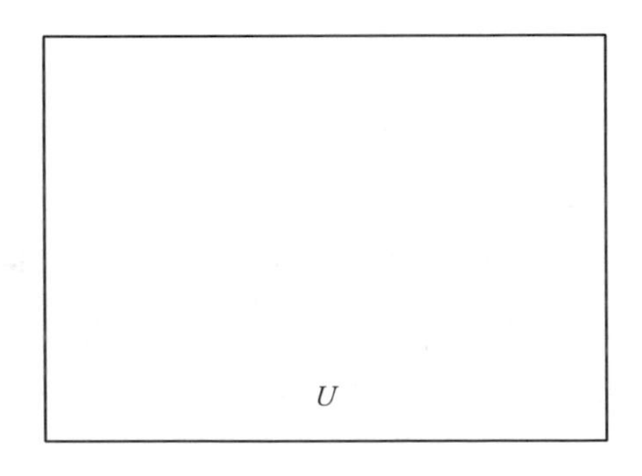

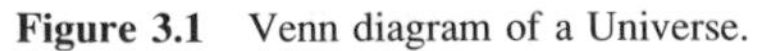
Figure 3.1 Venn diagram of a Universe.

[1] The existence of sets containing an infinite number of elements caused considerable controversy in the early days of the mathematical field of *set theory*. The paradoxes and logical contradictions that infinite sets give rise to were resolved only with much difficulty over several decades.

[2] Consider, for example, the famous Russell paradox, named for the philosopher/logician Bertrand Russell: Is the set of all sets which are not elements of themselves an element of itself?

Any set we wish to talk about (other than the Universe itself) is represented in a Venn diagram by a closed shape (often a circle) inside the Universe. Elements of the set are assumed to lie inside the shape; objects that are not elements of the set are assumed to lie outside the shape. For example, Figure 3.2 is a Venn diagram of a Universe containing one set.

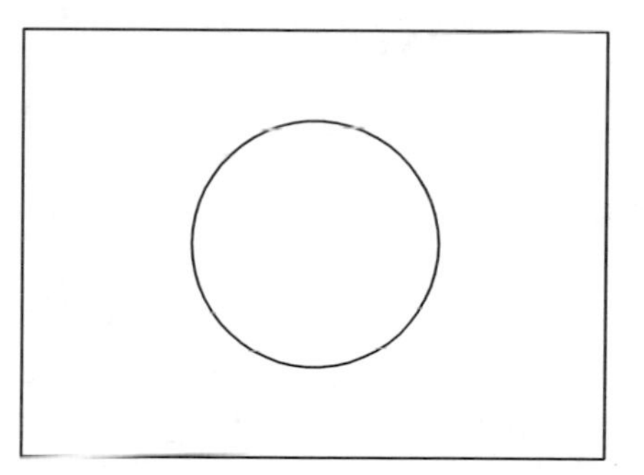

Figure 3.2 A Universe containing a set.

Call the set pictured in Figure 3.2 set x. In a Venn diagram a particular set is indicated by shading it in. In complex Venn diagrams, there may be several shaded areas, all of which collectively constitute a set. To indicate the set x itself and all elements therein, we shade in the circle representing x as in Figure 3.3.

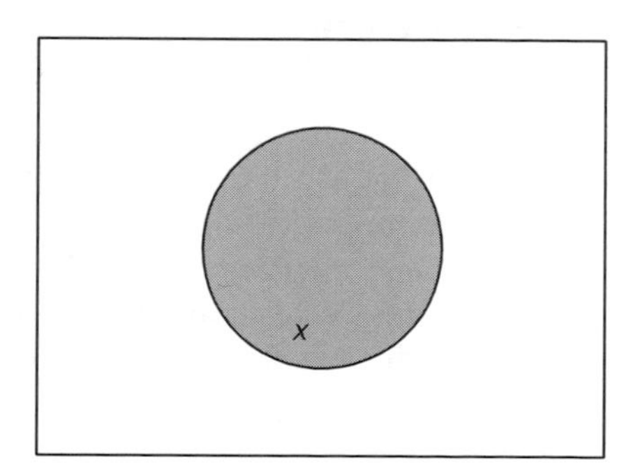

Figure 3.3 Venn diagram of the set x.

3.3 SET COMPLEMENTATION

To draw a Venn diagram of all objects in the Universe that are not in x, we begin with the diagram of the Universe containing the set x and we shade in everything in the Universe that is not inside x as in Figure 3.4.

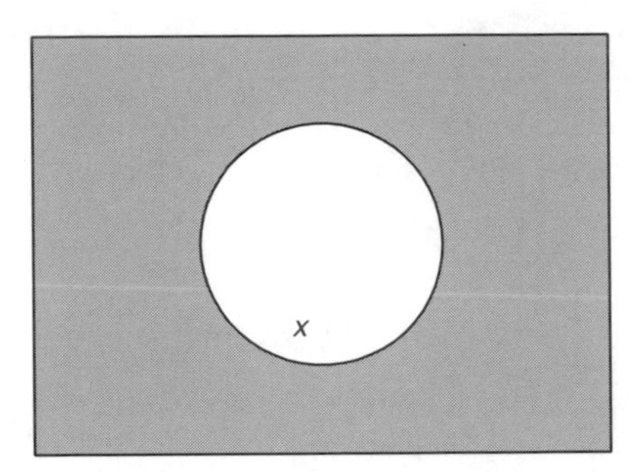

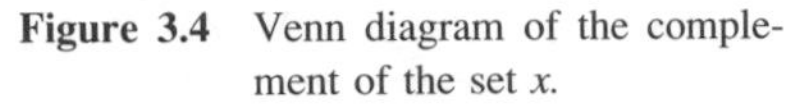
Figure 3.4 Venn diagram of the complement of the set x.

This is a set, too. It is the set of all objects that are not in x. If x is the set of sharp objects, then this is the set of nonsharp objects. If x is the set of blue dumpsters, then this is the set of everything in the Universe except blue dumpsters. We call this the complement of x or x' for short. Set complementation is the first of the functions that we will learn that operate within the algebra of sets.

As you have probably guessed by now, this concept of set complementation is analogous to the Boolean complementation function, NOT. Thus, the Venn diagram of set x corresponds to the Boolean bit x, and the Venn diagram of the set x' corresponds to the Boolean bit $\bar{x}$. Before we move on to the other basic functions of the algebra of sets, however, there are two more concepts that we should discuss.

3.4 THE NULL SET

The Universe is itself a set, that which contains everything. What, then, is its complement? In the algebra of sets, there is a concept known as the **null set.** Simply put, the null set is the imaginary set that contains nothing. It has no elements in it, and to represent it, we leave the entire Venn diagram unshaded (as opposed to the Universe, for which we shade in the entire Venn diagram). The symbol for the null set is $\varnothing$. Therefore $\varnothing' = U$ and, of course, $U' = \varnothing$.

3.5 SUBSETS AND SUPERSETS

The related concepts of **subset** and **superset** describe the relationships between sets. When all of the elements of one set are also elements of another set, the first set is called a subset of the second set, and the second set is called a superset of the first set. If set p is a subset of set q, then any element of set p is also an element of set q. The symbol $\subset$ means "is a subset of." Thus, we write $p \subset q$ to mean "set p

is a subset of set q'', or equivalently, ''set q is a superset of set p.''

For example, the set of all Alaskans is a subset of the set of all people. Any element of the set of Alaskans (i.e., a person who happens to live in Alaska) is also automatically a person. Likewise, the set of all people is a subset of the set of all mammals (and, syllogistically, the set of all Alaskans is also a subset of the set of all mammals). For the purposes of the algebra of sets, any set is considered to be a subset of itself, and the null set is a subset of every set.

3.6 INTERSECTION

Consider the set of all red things and the set of all round things. There are some things that are in one set but definitely outside the other, such as a fire engine (red, but not round) or a baseball (round but not red). But there are some things that are in both sets at the same time, such as an apple, which is both round and red. To draw a Venn diagram representing this situation, we need a Universe with two sets in it. The sets must be drawn in such a way that they overlap. The objects that are both round and red lie in this overlapping section (Fig. 3.5).

The shaded area constitutes a new set, the set of things that are both round and red. This new set is called the **intersection** of the two original sets. Thus, the intersection of two (or more) sets is the set of objects that lie in both (or all)

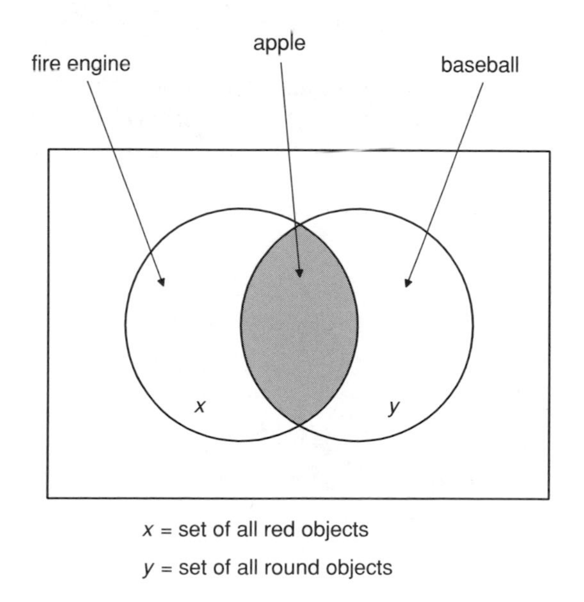

x = set of all red objects

y = set of all round objects

Figure 3.5 The set of all round, red objects.

of those sets, just as the intersection of two roads is that patch of pavement that is simultaneously part of both roads. Whenever we draw Venn diagrams of more than one set, we must draw all sets so that they overlap so that we can shade the intersection of any of the sets in the diagram.

Intersection is an operation on two or more sets that results in another set and as such is the second of our functions in the algebra of sets. It should be noted that the intersection of two or more sets is always a subset of each of the intersecting sets. The symbol for intersection looks like this: $\cap$. Thus, the intersection of sets a and b is denoted $a \cap b$ (Fig. 3.6).

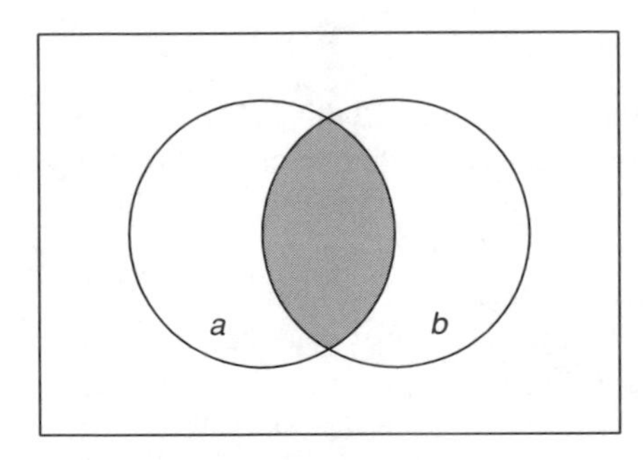

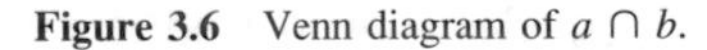
Figure 3.6 Venn diagram of $a \cap b$.

What is the intersection of the set of all nineteenth century German philosophers and the set of all kitchen appliances? Their intersection exists, but because there are no things that are both nineteenth century German philosophers and kitchen appliances, the intersection has no elements in it. It is therefore the null set.

Let us look at a Venn diagram of the intersection of three sets. To lie within this intersection, an object would have to be an element of all three sets. First, we have to draw the Universe with the three sets in it. Keep in mind that all the sets must overlap each other in the diagram (Fig. 3.7).

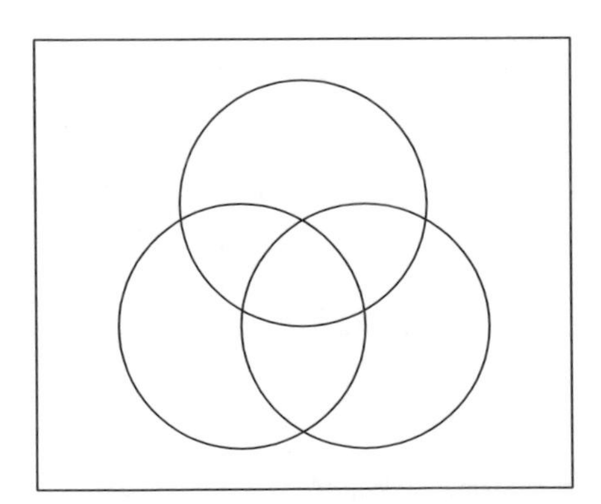

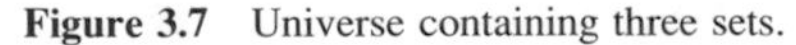
Figure 3.7 Universe containing three sets.

The intersection is the middle part that is in all three sets simultaneously (Fig. 3.8).

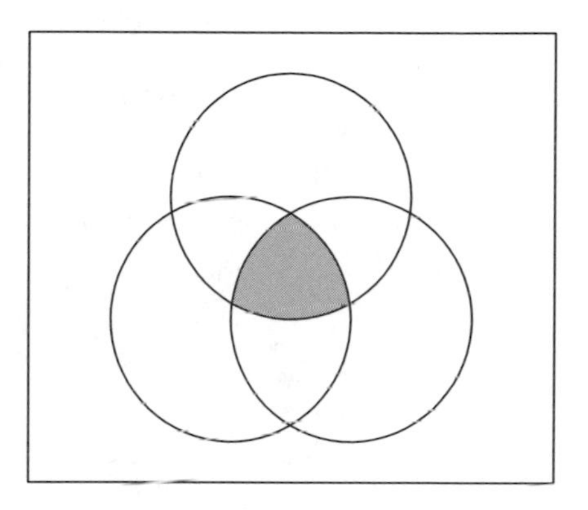

Figure 3.8 Intersection of three sets.

The intersection function in the algebra of sets is analogous to the Boolean function AND. To be in the intersection of two sets, an object must be an element of the first set AND an element of the second. If an object falls outside of even one of the sets being intersected, it is not in their intersection.

3.7 UNION

The third function in the algebra of sets that we will study is called **union.** The word "union" implies a merging of separate things. In the context of the algebra of sets, a union of two or more sets is the set formed by merging all of the sets in the union. Union is analogous to the Boolean function OR. To be in the union of any two sets, an object only has to be an element of one of the sets OR the other (or both). The only way an element can fail to be in the union of a number of sets is for it not to be an element of any of the sets in the union. A set is always a subset of the union of itself and any other set(s). The symbol for union is $\cup$.

The Venn diagram of the union of two sets x and y is shown in Figure 3.9.

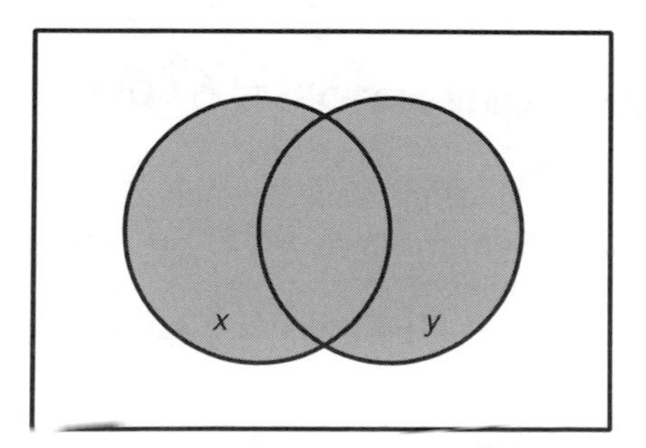

Figure 3.9 Venn diagram of $x \cup y$.

3.8 EXAMPLE OF UNION AND INTERSECTION

Suppose you were to draft a football team, and for your team you want people who are fast or big or both. The set of people eligible for your team would be the union of the set of fast people and the set of big people. This union would include the fast small people, the big slow people, and the big fast people. In fact, the only people excluded from the pool of eligible players would be the people who were not members of either set, i.e., the people who are small and slow (Fig. 3.10).

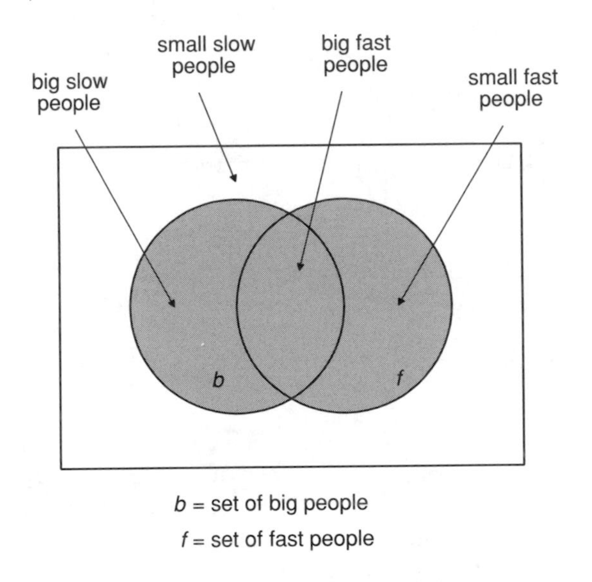

b = set of big people

f = set of fast people

Figure 3.10 Shaded area = eligible players.

Now suppose that your standards were a little higher for your team, and you only wanted people who were both fast AND big. The set of eligible players would consist of the intersection of the set of fast people and the set of big people (Fig. 3.11).

3.9 COMBINATORICS OF VENN DIAGRAMS

Look again at the Venn diagram of sets of fast and big people (Fig. 3.10). There are four distinct areas within the diagram, representing the sets of small slow people, small fast people, big slow people, and big fast people. These four categories represent all possible combinations of big/small and slow/fast. The intersection of the two sets only includes the last of the four combinations (big and fast), while their union includes all but the first, the set of small slow people. Table

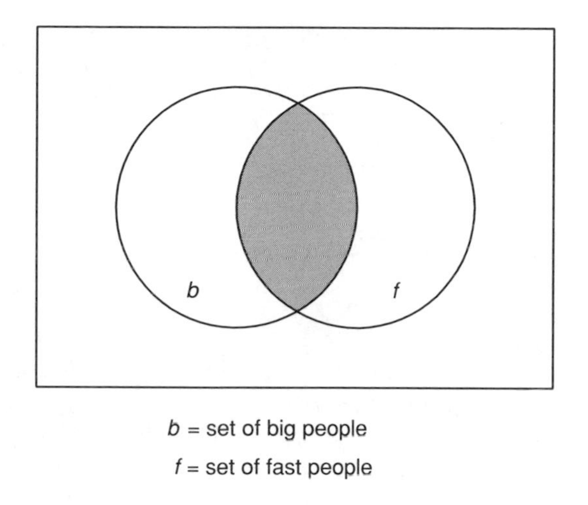

b = set of big people

f = set of fast people

Figure 3.11 Shaded area = eligible players, i.e., those who are both fast and big.

3.1 is a chart of all combinations of big/small and slow/fast people. The table shows which kinds of people are inside the union and which kinds of people are inside the intersection of the sets of big people and fast people.

TABLE 3.1

Kind of Person	In Set: Big ∪ Fast?	In Set: Big ∩ Fast?
small slow	no	no
small fast	yes	no
big slow	yes	no
big fast	yes	yes

This table illustrates clearly why union is analogous to OR and why intersection is analogous to AND. Their truth tables are the same, just in a slightly different syntax (no = 0, yes = 1). Indeed, we can create a more traditional truth table that represents the same situation. The first input bit is called *big* (1 = true, i.e., big, 0 = false, i.e. small) and the second input bit *fast* (1 = true, i.e., fast, 0 = false, i.e., slow) (Table 3.2).

It should be apparent why it is important that we draw all sets in Venn diagrams in such a way that they all overlap. Specifically, as we add each new set to a Venn diagram that we are drawing, the boundary of the new set must pass through every closed area already in the diagram. This way, the new set bisects each existing area, thereby doubling the total number of distinct closed areas in the diagram and ensuring that we have areas in the diagram to represent all possible combinations of the sets in the Universe without leaving any combinations out.

TABLE 3.2

Big	Fast	Big $\vee$ Fast	Big $\wedge$ Fast
0	0	0	0
0	1	1	0
1	0	1	0
1	1	1	1

However many sets are shown in a Venn diagram, the diagram should have two to the power of that number of distinct closed areas. In a diagram of two sets, there are 2^2 or four distinct areas. In a Venn diagram of three sets, there are 2^3 = eight distinct areas. In general, a Venn diagram of n sets should have 2^n distinct closed areas in it. In practice, however, Venn diagrams with more than three sets are tricky and rarely used.

3.10 NUMBERING REGIONS IN VENN DIAGRAMS

The truth table for a Boolean expression with n bits has 2^n rows in it; a Venn diagram containing n sets has 2^n distinct areas in it. There is one area in the Venn diagram for each row in the corresponding truth table. We can number the areas in a Venn diagram in much the same way that we number the rows in a truth table in base 2. Let us number the areas in a Venn diagram containing three sets. Figure 3.12 shows its blank Venn diagram.

Now we must establish an arbitrary order. Because we named the sets x, y, and z, we will use alphabetical order, x, y, z. We will assign a different three-bit number to each of the eight distinct areas in the Venn diagram. The individual bits in these numbers will be called x, y, and z. For each of the eight areas the

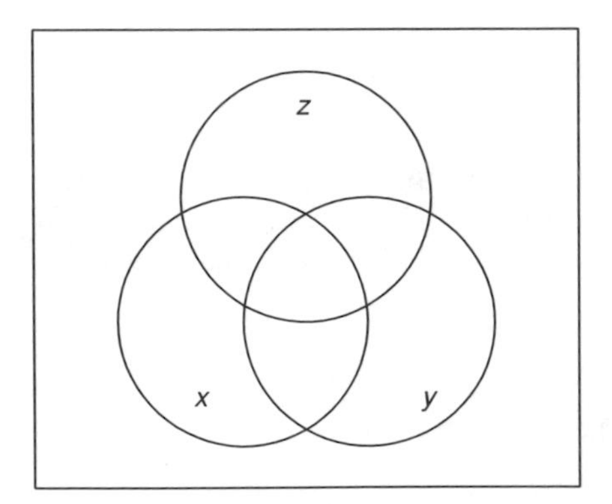

Figure 3.12 Three sets.

value of each bit will indicate that area's inclusion in or exclusion from the set for which the bit is named. If a given area is inside set x, for example, the x bit will be on (1) in that area's number. If it is outside x, the x bit will be off (0). Likewise for sets (and bits) y and z. This way we can count the areas in the Venn diagram, assigning to each a unique number in the range 000_2–111_2 (0_{10}–7_{10}). Just as the range of numbers from 000_2 to 111_2 represents all possible combinations of three bits, a correctly drawn Venn diagram represents all possible combinations of inclusion in or exclusion from the sets depicted.

As an example, look at the area outside the sets in a three-set Venn diagram (Fig. 3.13).

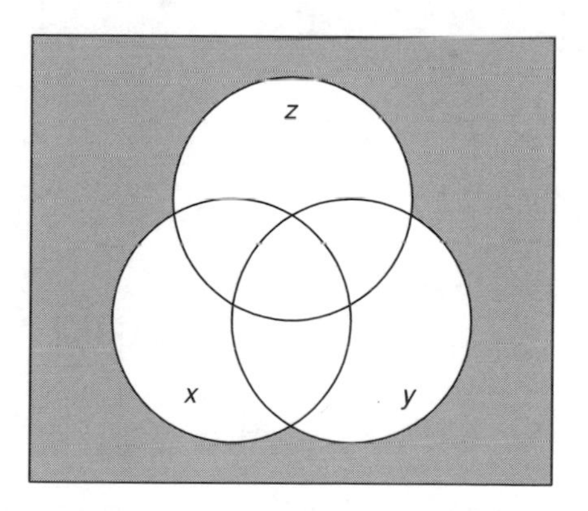

Figure 3.13 Area outside all three sets (area 0).

Because this area is outside all three sets, x, y, and z, all three bits in its number will be zero: $(x, y, z) = (000)$. So this area's number is $000_2 = 0$. Now consider the area that is the intersection of all three sets (Fig. 3.14).

Because this area is inside all three sets, all three bits in its number are 1: $(x, y, z) = (111)$, so this area's number is $111_2 = 7$. Consider the area in Figure 3.15.

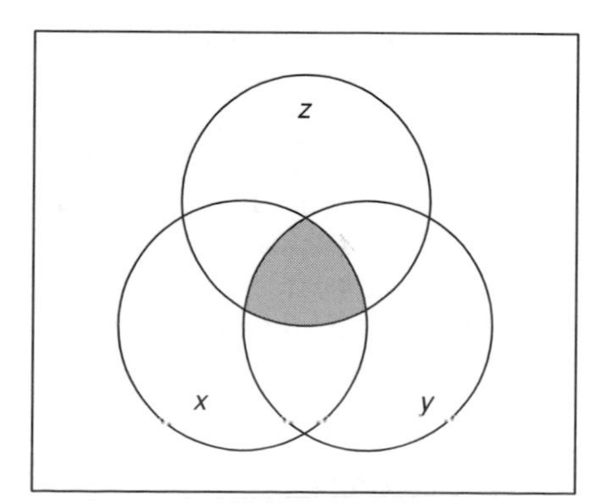

Figure 3.14 Intersection of three sets (area 7).

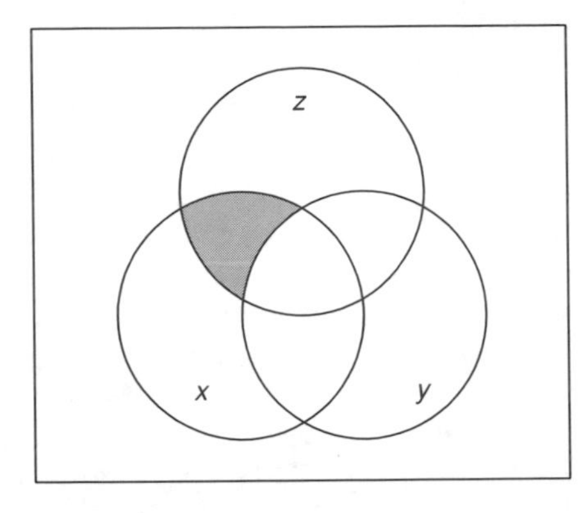

Figure 3.15 Area 5.

This area is inside of x, so $x = 1$. However, it is outside of y, so $y = 0$. Finally, it is inside of z, so $z = 1$. For this area, then, $(x, y, z) = (101) = 5_{10}$.

EXERCISE 3.1

1. Draw a Venn diagram containing three sets, x, y, and z. Number each of the regions and shade in region 3.
2. Draw a four-set Venn diagram. (This is a trick question. The drawing you are most likely to draw simply does not work—it is nearly useless for illustrating functions in the algebra of sets. Explain why and try to draw a valid four-set Venn diagram.)

3.11 COMBINATIONAL LOGIC IN VENN DIAGRAMS

For any Boolean expression with n bits (where $n \leq 3$), in addition to writing a truth table, we can draw a Venn diagram. This diagram will have n overlapping sets and the shaded areas in the diagram will correspond to the rows in the truth table in which there is a 1 in the output column. One way to draw such a diagram is simply to shade in the areas in the Venn diagram whose numbers correspond to those rows in the truth table which have 1's in the output column. For example, consider the two-bit expression $x \vee \bar{y}$. Its truth table is shown in Table 3.3.

The Venn diagram will have two sets, x and y. The four distinct areas of the Venn diagram are numbered as shown in Figure 3.16.

Since there are 1's in the output columns of the truth table in the rows in which $(x, y) = (0, 0)$, $(1, 0)$, and $(1, 1)$, we shade in these areas in the Venn diagram is shown in Figure 3.17.

TABLE 3.3

x	y	$x \vee \bar{y}$
0	0	1
0	1	0
1	0	1
1	1	1

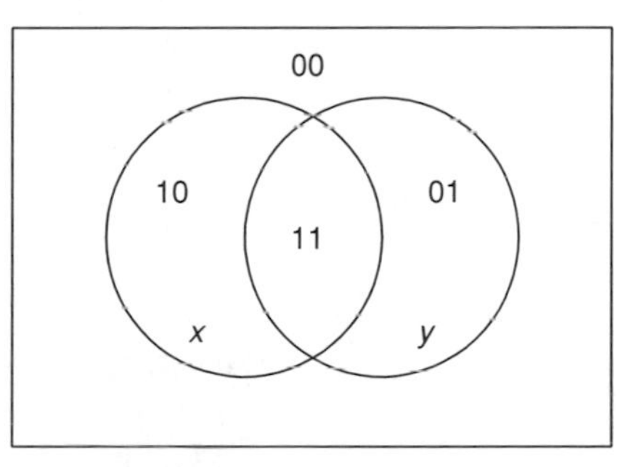

Figure 3.16 Two sets with all areas numbered.

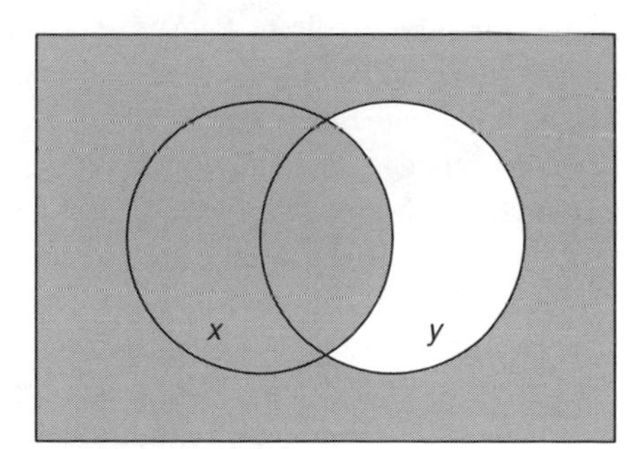

Figure 3.17 Venn diagram of $x \cup y'$.

3.12 SET ALGEBRAIC INTERPRETATION OF COMBINATIONAL LOGIC

That method, however, is a somewhat mechanical way of representing a Boolean function. We are just "stealing" the information from the truth table without interpreting it in any way. Used in this way, Venn diagrams are just a way of drawing pictures of truth tables. We will now use our understanding of the concepts of the algebra of sets to interpret a little more intelligently the Boolean function $x \vee \bar{y}$ in a Venn diagram. We do this by reinterpreting the Boolean function in terms of the algebra of sets. This is a notational change, from $x \vee \bar{y}$

to $x \cup y'$. Then we draw the Venn diagram of the resulting function in the algebra of sets.

$x \cup y'$ is the union of two sets, x and y' in a Universe that contains only those two sets (Fig. 3.18).

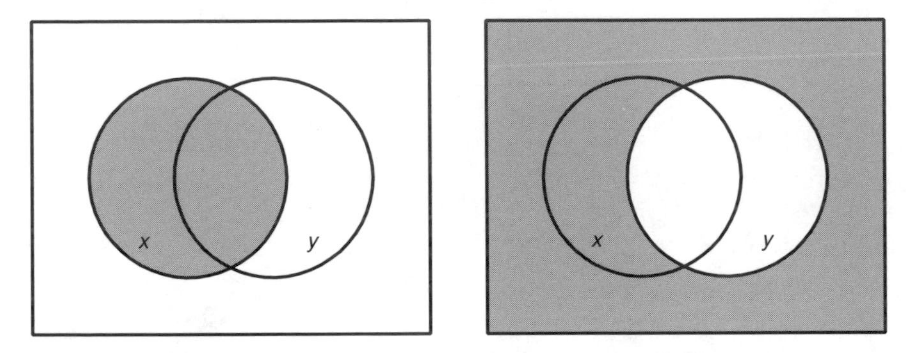

Figure 3.18 Two-set Venn diagrams representing the sets x and y', respectively.

To draw the union of these two sets, we shade in the parts of the diagram that are shaded in either or both of the diagrams of x and y' is shown in Figure 3.19.

This is the same Venn diagram we drew before. We just interpreted things a little differently this time. As we saw in chapter 2, Boolean algebra is a mathematical system with things called bits and the functions AND, OR, and NOT, which operate on the bits. Venn's algebra of sets is another mathematical system with things called sets and the functions union, intersection, and complementation that operate on the sets. At first these two systems appear to be different, but by now it should be clear that they are not simply similar, they are in a sense exactly the same mathematical system expressed in a different language and conceptualized in different ways.

Corresponding to every Boolean function for which we can write a truth table, algebraic expression, or circuit diagram, there is a set that we can describe using the symbolic language of the algebra of sets. The reverse is also true. When

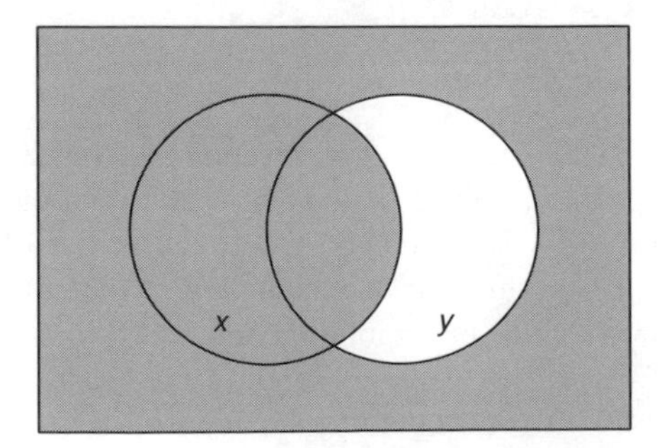

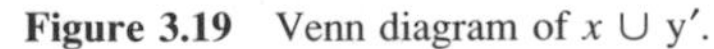

Figure 3.19 Venn diagram of $x \cup y'$.

two mathematical systems can be shown to be different ways of expressing exactly the same concepts in this way, such that there is an explicit one-to-one correspondence between statements in one system and statements in the other, an **isomorphism** is said to exist between the systems. This word comes from the Greek for "same body" or "same shape." We say that Boolean algebra is isomorphic to the algebra of sets.[3]

For a more involved example, consider the three-bit Boolean expression $(x \wedge \bar{y}) \vee \overline{(x \vee z)}$. Its truth table is shown in Table 3.4.

The blank Venn diagram, with all areas numbered, looks like Figure 3.20.

TABLE 3.4

x	y	z	$(x \wedge \bar{y}) \vee \overline{(x \vee z)}$
0	0	0	1
0	0	1	0
0	1	0	1
0	1	1	0
1	0	0	1
1	0	1	1
1	1	0	0
1	1	1	0

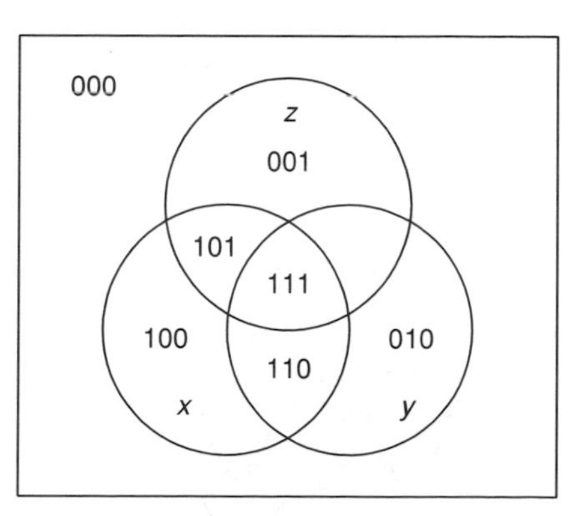

Figure 3.20 Three sets with all areas numbered.

[3] Boolean algebra refers not just to one mathematical system, but to a whole family of mathematical systems, the different varieties of which are beyond the scope of this book. However, in 1934 an American mathematician named Marshall Stone published the Stone Representation Theorem which proved that all Boolean algebras are isomorphic with some mathematical system based on sets.

If we just shade in the areas in the Venn diagram that have numbers that correspond to rows in the truth table in which there is a 1 in the output column, (areas 000, 010, 100, and 101) we get the diagram shown in Figure 3.21.

The Venn diagram of the set $(x \cap y') \cup (x \cup z)'$ is the union of the sets $x \cap y'$ and $(x \cup z)'$, which are themselves functions of other sets. To draw $x \cap y'$, we must first draw the sets of x and y', keeping in mind that since we have a three-set function, we must include z in our Universe, even though we will not use it until later (Fig. 3.22).

We now draw their intersection $x \cap y'$ by drawing a Venn diagram (Fig. 3.23) in which we shade only those areas that are shaded in Figures 3.21 and 3.22.

Next, we need to draw the Venn diagram of $(x \cup z)'$. This is the complement

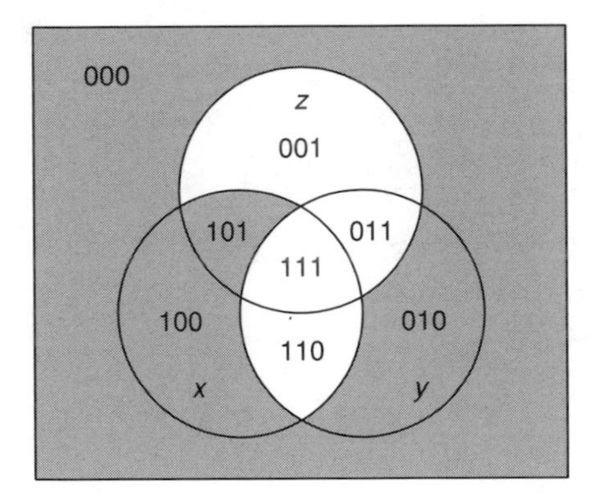

Figure 3.21 $(x \cap y') \cup (x \cup z)'$.

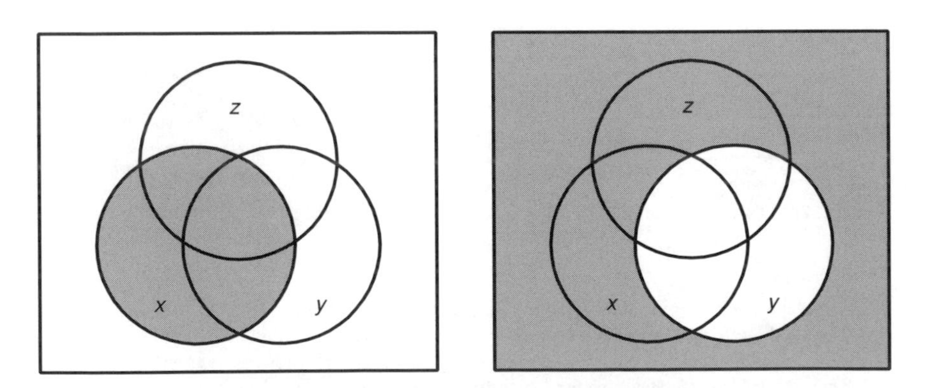

Figure 3.22 Three-set Venn diagrams of the sets x and y', respectively.

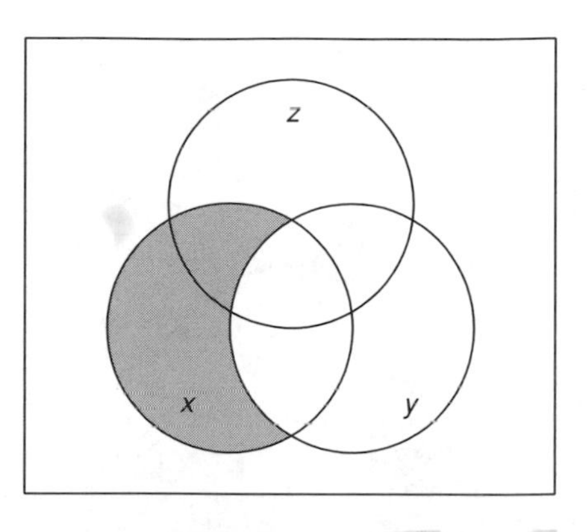

Figure 3.23 Three-set Venn diagram of $x \cap y'$.

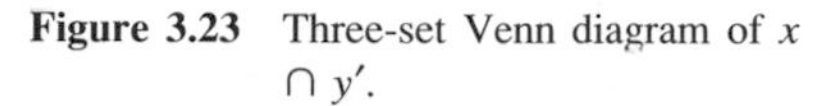

of the union of x and z. We already know how to draw the union of two sets as shown in Figure 3.24.

To draw $(x \cup z)'$, we draw the complement of the union, shown in Figure 3.24, shading in the entire Venn diagram except that portion shaded in the diagram of $x \cup y$ (Fig. 3.25).

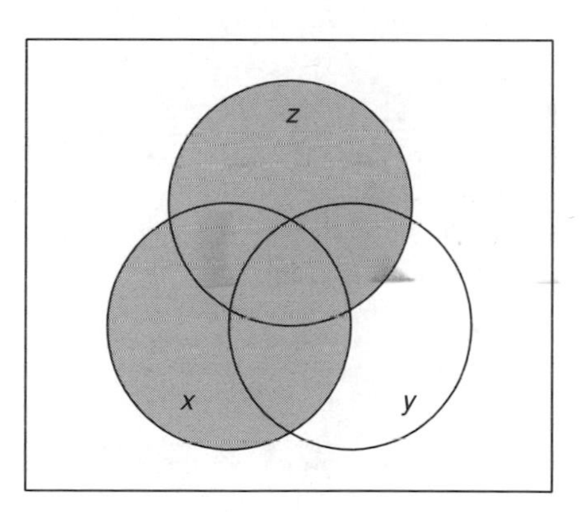

Figure 3.24 Three-set Venn diagram of $x \cup z$.

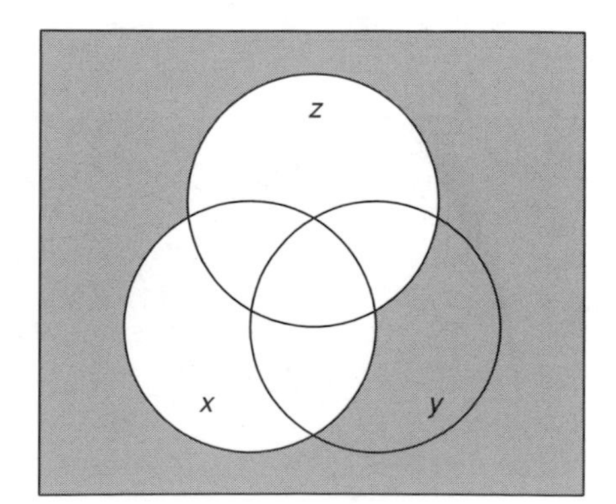

Figure 3.25 Three-set Venn diagram of $(x \cup z)'$.

Finally, we draw the Venn diagram of $(x \cap y') \cup (x \cup z)'$ by drawing the union of the sets $x \cap y'$ and $(x \cup z)'$ (Fig. 3.26). We do this by shading in all areas that are shaded in either or both of the diagrams we drew for $(x \cap y')$ (Fig. 3.23) and $(x \cup z)'$ (Fig. 3.25).

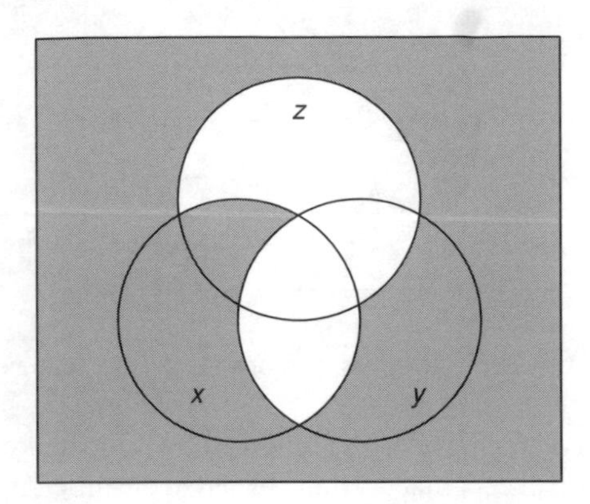

Figure 3.26 $(x \cap y') \cup (x \cup z)'$.

This is the same as the Venn diagram we "stole" directly from the truth table for the Boolean algebraic expression $(x \wedge \bar{y}) \vee \overline{(x \vee z)}$ (Fig. 3.21).

EXERCISE 3.2

A. For each of the following Boolean expressions: (a) write the truth and (b) draw the Venn diagram by numbering each region in the blank Venn diagram and "stealing" directly from the truth table. Check your work by drawing the Venn diagram of the set algebraic expression corresponding to the given Boolean expression, showing all intermediate Venn diagrams.

1. $x \vee y$
2. $(x \vee \bar{y} \vee \bar{z}) \wedge \overline{(\bar{x} \wedge y \wedge \bar{z})}$
3. $\bar{x} \vee y$
4. $\overline{(x \wedge \bar{y})}$
5. $(x \vee y \vee z) \wedge (x \vee \bar{y} \vee \bar{z}) \wedge (\bar{x} \vee y \vee \bar{z}) \wedge (\bar{x} \vee \bar{y} \wedge z)$
6. $(\bar{a} \wedge \bar{b} \wedge c) \vee (\bar{a} \wedge b \wedge \bar{c}) \vee (a \wedge \bar{b} \wedge \bar{c}) \vee (a \wedge b \wedge c)$
7. $(x \vee y \vee z) \wedge (x \wedge y \wedge \bar{z}) \wedge \overline{(x \wedge \bar{y})}$
8. $\overline{(x \wedge y)} \vee x$

4

Other Boolean Functions

Now that we have some knowledge of the algebra of sets and the basics of combinational logic, it is time to expand our repertoire of Boolean functions. We know about the two-input AND and OR functions, but what other two-input Boolean functions might there be? Any such function can be specified by a truth table with two input columns and one output column, and $2^2 = 4$ rows (Table 4.1).

Because each two-input Boolean function is completely characterized by the values of the bits in the output column of its truth table, what we are really asking is how many different ways are there of filling in the output column of the truth table in Table 4.1. Since there are four bits in the output column, the question becomes how many different combinations of four bits there are, which we know is 2^4 or 16: 0000 (0_{10}) through 1111 (15_{10}). Rather than write out all 16 truth tables to represent all 16 different two-bit functions, we will write one big truth table with 16 output columns, one for each possible two-bit function (Table 4.2). Each output column is numbered with the base 10 equivalent of the bits that occupy it (for the purposes of this numbering scheme, each output column is considered to be a four-bit binary number with its most significant bit at the bottom). In this way, the output columns count themselves from 0000 at the far left to 1111 at the far right.

Any possible two-input Boolean function is represented by one of the 16 output columns in the truth table shown in Table 4.2; AND is represented by column eight and OR by column 14. Of the 14 two-bit functions besides AND and OR, some are not terribly interesting, but others are worth special attention.

TABLE 4.1

x	y	Output
0	0	?
0	1	?
1	0	?
1	1	?

TABLE 4.2 Output Columns of the Truth Tables for All Possible Two-input Boolean Functions

x	y	0	1	2	3	4	5	6	7	8	9	10	11	12	13	14	15
0	0	0	1	0	1	0	1	0	1	0	1	0	1	0	1	0	1
0	1	0	0	1	1	0	0	1	1	0	0	1	1	0	0	1	1
1	0	0	0	0	0	1	1	1	1	0	0	0	0	1	1	1	1
1	1	0	0	0	0	0	0	0	0	1	1	1	1	1	1	1	1

However, we will find that all of them can be constructed (or realized) with the right combination of ANDs, ORs and NOTs.

4.1 THE CONSTANT FUNCTIONS 0 AND 1

In Table 4.2, output column 0 has nothing but 0's in all four rows, while output column 15 contains nothing but 1's. These two functions are not really functions at all in the sense that all of the others are functions. They completely ignore their inputs. No matter what its input bits are, the function specified by column 0 will always produce a 0 as output, and the function specified by column 15 will always produce a 1. Because their outputs are constant, these are called the **constant 0 function** and the **constant 1 function,** respectively. We have seen examples of constant functions in previous chapters, and we know that any function that turns out to be the constant 1 function is a tautology. There are no circuit diagrams for the constant functions 0 and 1.

The Venn diagrams corresponding to the two-input constant 0 function and the constant 1 function are simple (Fig. 4.1). Both functions have two input bits, so the Venn diagrams contain two overlapping sets. For the constant 0 function, none of the diagram is shaded. The set represented by 0 is, in fact, the null set. For the constant 1 function, all of the diagram is shaded. The equivalent to the Boolean constant 1 function in the algebra of sets is the entire Universe.

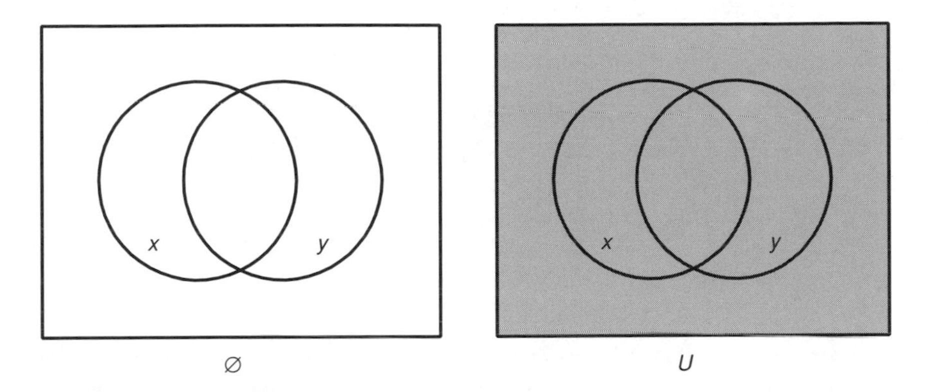

Figure 4.1 Venn diagrams corresponding to the constant 0 function and the constant 1 function.

4.2 NAND

The constant functions 0 and 1 have their uses as we shall see, but because they do not really do anything (they are constant, after all) they are not very interesting. We will turn our attention to the function represented by column seven in Table 4.2 whose truth table is shown in Table 4.3.

TABLE 4.3

x	y	Function 7
0	0	1
0	1	1
1	0	1
1	1	0

This function is 1 whenever either (or both) of its inputs is 0, and 0 only when both of its inputs are 1. A comparison of the output column of this function with that of the two-input AND function reveals that they are complementary: their output bits differ in each of the four rows of their truth tables (Table 4.4).

For this reason, this function is called the NAND function, short for NOT AND. It is most often represented algebraically according to this understanding of it: $\overline{(x \wedge y)}$. However, there is a special algebraic symbol for NAND. It is a vertical bar (|), although to distinguish it more clearly from the Peirce arrow, which we will see later, it is sometimes rendered as an up arrow (↑). We will use

TABLE 4.4

x	y	Function 7	$x \wedge y$
0	0	1	0
0	1	1	0
1	0	1	0
1	1	0	1

this latter notation,[1] so $x \uparrow y = \overline{(x \wedge y)}$. This NAND symbol is called a **Sheffer stroke** (and NAND is likewise sometimes called the Sheffer stroke function) after Henry M. Sheffer (1883–1964). Although Sheffer established the algebraic notation and explored some of NAND's properties in a paper published in 1913, much of his work was anticipated years before by C. S. Peirce (see NOR below).

NAND has no real circuit diagram of its own, but we do call an AND gate with a NOT bubble on its output line a NAND gate. This construction is shown in Figure 4.2.

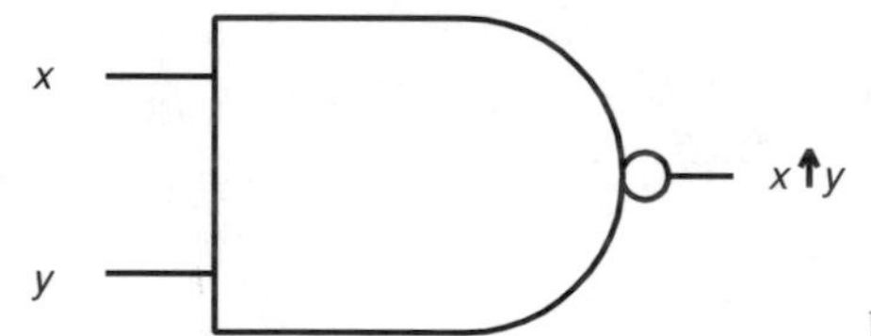

Figure 4.2 NAND gate.

Since NAND is the exact complement of AND, and the intersection function is the Venn diagram analogue of the Boolean AND function, the function corresponding to NAND is the algebra of sets is the complement of the intersection function, or $(x \cap y)'$ (Figure 4.3).

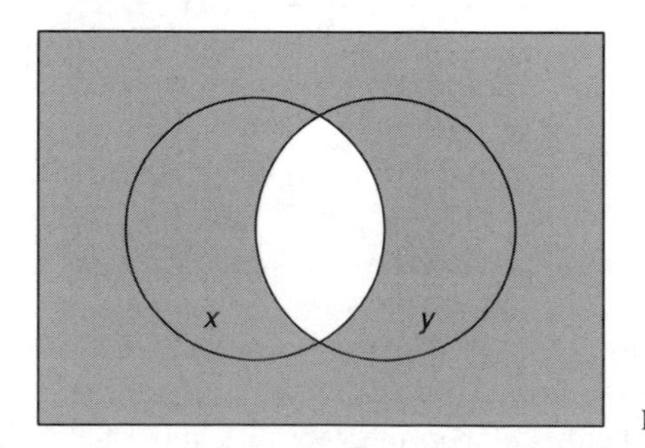

Figure 4.3 $(x \cap y)'$.

[1] In some contexts, notably certain computer programming languages, a vertical bar is taken to mean Boolean OR, giving us even more reason to use the up arrow to represent NAND.

4.3 NOR

You may have already guessed that NOR stands for NOT OR. The NOR function is represented by column 1 in Table 4.2. It is the complement of the OR function; it is 1 only when both of its inputs are 0, and 0 if either (or both) of its inputs are 1. NOR's truth table is given in Table 4.5.

TABLE 4.5

x	y	NOR (function 1)	$x \vee y$
0	0	1	0
0	1	0	1
1	0	0	1
1	1	0	1

Just as we most often represent the NAND function in terms of AND and NOT, we will most often represent NOR in terms of OR and NOT: $\overline{(x \vee y)}$. However, there is a special algebraic symbol for NOR as well as for NAND. It is a single down arrow: $\downarrow$. Thus $x \downarrow y = \overline{(x \vee y)}$. Used to represent NOR, this symbol is called a **Peirce Arrow** after Charles Sanders Peirce.[2]

As with the NAND function, there is no unique circuit symbol for NOR, just an OR gate with a NOT bubble on its output line as seen in Figure 4.4.

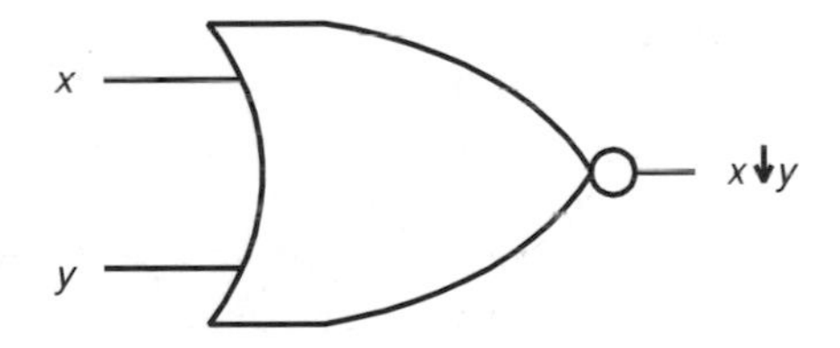

Figure 4.4 NOR gate.

The set algebraic function corresponding to NOR is the complement of the union of two sets, $(x \cup y)'$. The Venn diagram of this function is shown in Figure 4.5.

4.4 XOR

The OR function we now use, sometimes called the inclusive OR function (or IOR for short), superseded Boole's original OR function in which 1 OR 1 was undefined. 1 OR 1 was simply not an allowed function, just as $1 \div 0$ is not

[2] Peirce is arguably the first designer of a digital circuit, and symbolic logic is just one of many fields in which he made significant contributions. Others include, but are by no means limited to, astronomy, physics, cartography, and mathematics. In spite of his achievements, however, he died in abject poverty.

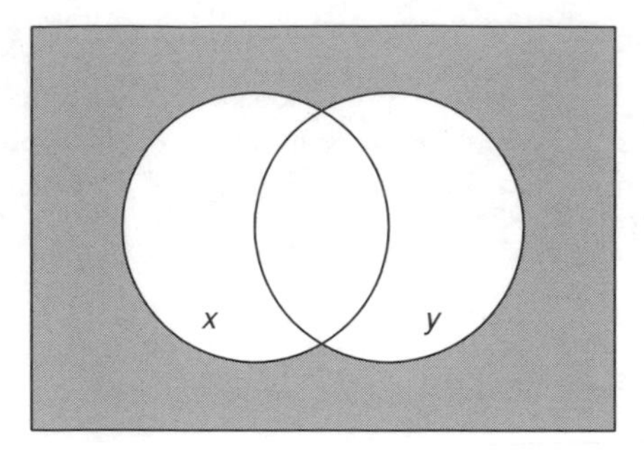

Figure 4.5 $(x \cup y)'$.

allowed in ordinary arithmetic. However, a form of Boole's original OR function is still used. It has been modified so that while the formerly undefined case is now defined, it is defined to be 0 rather than 1. Thus, in a sense, this descendent of Boole's OR function still excludes the case in which both inputs are 1 and is accordingly called **exclusive OR** (or **XOR,** pronounced "ex-or"). 1 XOR 1 = 0, but in all other cases XOR acts exactly like the inclusive OR function. The XOR function is represented by column six of Table 4.2. Its truth table is also shown in Table 4.6. (OR is shown as well for comparison).

XOR has a unique algebraic symbol. It consists of a circle with a plus sign through it: $\oplus$. Unlike the symbols for NAND and NOR, we will use the XOR symbol often for convenience's sake, because it is somewhat complicated to represent the exclusive OR function in terms of AND, inclusive OR, and NOT alone. Nevertheless, we will now try to do so.

When is the XOR function equal to 1? According to the truth table, it is equal to 1 in two cases, represented by rows 1 and 2 of the truth table (remembering that we start with 0 when we number the rows of truth tables, so these are the second and third rows): 1) when $x = 0$ and $y = 1$, and 2) when $x = 1$ and $y = 0$. If the inputs to the XOR function conform to either of these two cases, the XOR function will be true (equal to 1). Worded slightly differently, x XOR y when ($x = 0$ AND $y = 1$) OR ($x = 1$ AND $y = 0$).

This sounds more like a Boolean algebra expression than an English sentence. Taking this notion a bit further, according to the principle of assertion,

TABLE 4.6

x	y	x XOR y	$x \vee y$
0	0	0	0
0	1	1	1
1	0	1	1
1	1	0	1

$x = 0$ is another way of saying $\bar{x}$, and $y = 1$ is another way of saying y. Thus ($x = 0$ AND $y = 1$) is equivalent to (NOT x AND y), or $\bar{x} \wedge y$. Consider the truth table for $\bar{x} \wedge y$ shown in Table 4.7.

TABLE 4.7

x	y	$\bar{x} \wedge y$
0	0	0
0	1	1
1	0	0
1	1	0

$\bar{x} \wedge y$ is true (equal to 1) when $x = 0$ and $y = 1$. Similarly, ($x = 1$ AND $y = 0$) is equivalent to the expression $x \wedge \bar{y}$ (which is 1 only when $(x, y) = (1, 0)$). So to say that x XOR y is 1 whenever ($x = 0$ AND $y = 1$) OR ($\bar{x} = 1$ AND $y = 0$) is the same as saying x XOR $y = (\bar{x} \wedge y) \vee (x \wedge \bar{y})$. To check the truth of this, look at the truth table for the function $(\bar{x} \wedge y) \vee (x \wedge \bar{y})$ in Table 4.8.

The output column is indeed the same as that of the XOR function, and we have shown that $(\bar{x} \wedge y) \vee (x \wedge \bar{y}) = x$ XOR y. It is now apparent why we will use $\oplus$ most of the time to represent XOR algebraically rather than expressing XOR in terms of AND, OR, and NOT as we did with NAND and NOR; $(\bar{x} \wedge y) \vee (x \wedge \bar{y})$ is somewhat less clear and convenient than $x \oplus y$.

TABLE 4.8

x	y	$\bar{x} \wedge y$	$x \wedge \bar{y}$	$(\bar{x} \wedge y) \vee (x \wedge \bar{y})$
0	0	0	0	0
0	1	1	0	1
1	0	0	1	1
1	1	0	0	0

The circuit diagram of XOR in terms of AND, OR, and NOT gates (Fig. 4.6) helps to clarify the XOR function. XOR does, however, have its own circuit symbol shown in Figure 4.7.

All that remains is to draw the Venn diagram corresponding to XOR. We could steal it directly from XOR's truth table, or we could draw the Venn diagram of the set algebraic equivalent of $(\bar{x} \wedge y) \vee (x \wedge \bar{y})$, namely $(x' \cap y) \cup (x \cap y')$. To do this, we would first draw the Venn diagrams of $x' \cap y$ and $x \cap y'$ and then draw the diagram of their union. Instead of either of these options, let us try a different approach. The only difference between the inclusive OR and

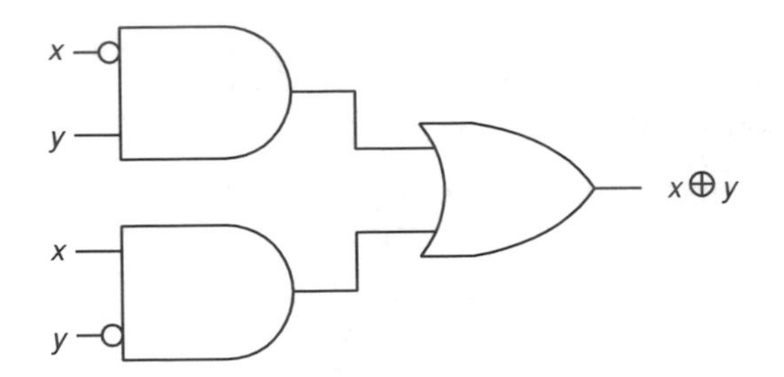

Figure 4.6 The XOR function realized in terms of AND, OR, and NOT.

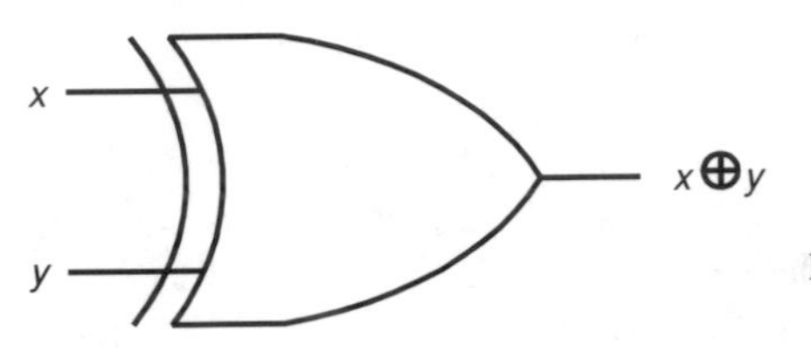

Figure 4.7 The circuit symbol of the XOR function.

XOR is that XOR does not include the case in which both inputs are 1. In that case, IOR is 1, but XOR is 0. Thus, the Venn diagram corresponding to XOR is exactly the same as that for the union function (OR) except that it is missing the part in the middle that is included in both sets (i.e., their intersection). XOR's Venn diagram is shown in Figure 4.8.

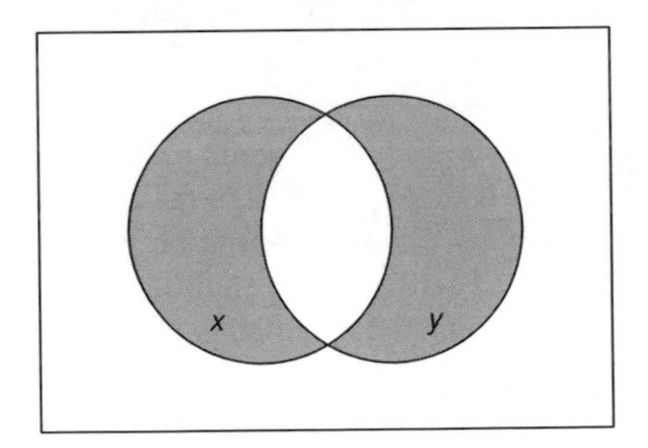

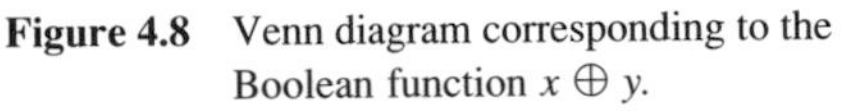

Figure 4.8 Venn diagram corresponding to the Boolean function $x \oplus y$.

4.5 COIN

The output of the **coincidence** function is given in column nine of Table 4.2 (and in the truth table of the coincidence function shown in Table 4.9). Coincidence is sometimes called the **COIN** function (pronounced ''co-in'', not like ''coin'' as in dime). It is called ***coincidence*** because it is 1 if and only if both of its inputs are the same, or when they **coincide.** That is, COIN is 1 when both inputs are 0 or when both inputs are 1, but it is 0 when one input is 0 and the other is 1.

TABLE 4.9

x	y	x COIN y
0	0	1
0	1	0
1	0	0
1	1	1

COIN has its own algebraic symbol, a circle with a dot in its center: $\odot$. Naturally, however, we will attempt to represent it in terms of AND, OR and NOT. Using the method we used above for XOR, we see that x COIN y when ($x = 0$ AND $y = 0$) OR ($x = 1$ AND $y = 1$). Translating this into terms of Boolean algebra by way of the principle of assertion, we have x COIN $y = (\bar{x} \wedge \bar{y}) \vee (x \wedge y)$. As with XOR, we can prove that this expression equals COIN by writing its truth table (Table 4.10).

TABLE 4.10

x	y	$\bar{x} \wedge \bar{y}$	$x \wedge y$	$(\bar{x} \wedge \bar{y}) \vee (x \wedge y)$
0	0	1	0	1
0	1	0	0	0
1	0	0	0	0
1	1	0	1	1

As expected, the final output column of the expression $(\bar{x} \wedge \bar{y}) \vee (x \wedge y)$ matches the output column for x COIN y exactly. Thus we have shown that $(\bar{x} \wedge \bar{y}) \vee (x \wedge y)$ is a realization of the coincidence function in terms of AND, OR, and NOT.

The circuit diagram for COIN, constructed in terms of AND, OR, and NOT gates, is shown in Figure 4.9.

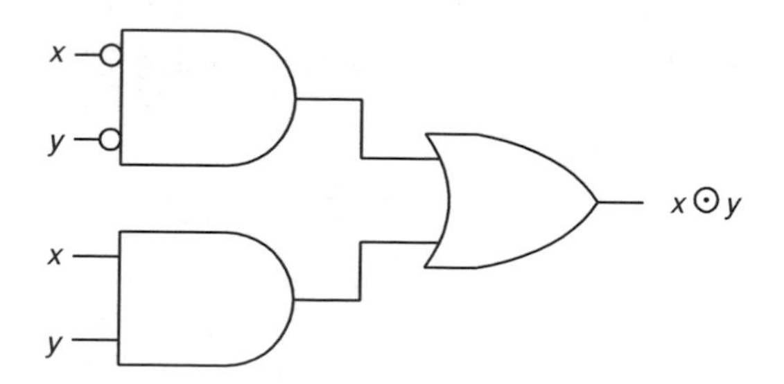

Figure 4.9 Circuit diagram of COIN realized in terms of AND, OR, and NOT.

Like XOR, however, COIN has its own circuit symbol, seen in Figure 4.10.

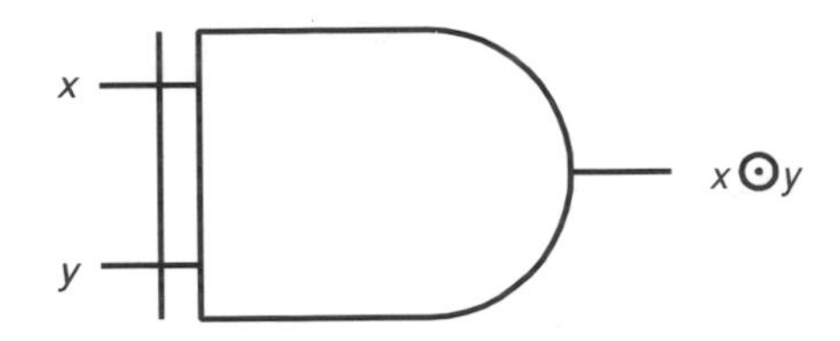

Figure 4.10 Circuit symbol of the COIN function.

4.5.1 Interesting Properties of XOR and COIN

Compare the truth tables of x XOR y and x COIN y in Table 4.11. They are complementary. Just as the coincidence function is 1 whenever both of its inputs are the same and 0 when they are different, the exclusive OR function is 1 whenever both of its inputs are different, and 0 when they are the same. For this reason, in electrical engineering circles the coincidence function is almost always called **exclusive NOR** (or XNOR for short, pronounced "ex-nor"), being the complement (NOT) of exclusive OR. We may realize either function in terms of the other and the NOT function, writing $x \oplus y = \overline{(x \odot y)}$ and $x \odot y = \overline{(x \oplus y)}$. This inverse relation between the XOR and COIN functions is illustrated by the circuit diagrams in Figures 4.11 and 4.12. In digital circuit diagrams used in electrical engineering schematics, however, the COIN function is most often

TABLE 4.11

x	y	$x \oplus y$	$x \odot y$
0	0	0	1
0	1	1	0
1	0	1	0
1	1	0	1

Figure 4.11 NOT COIN is equivalent to XOR.

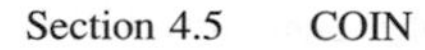

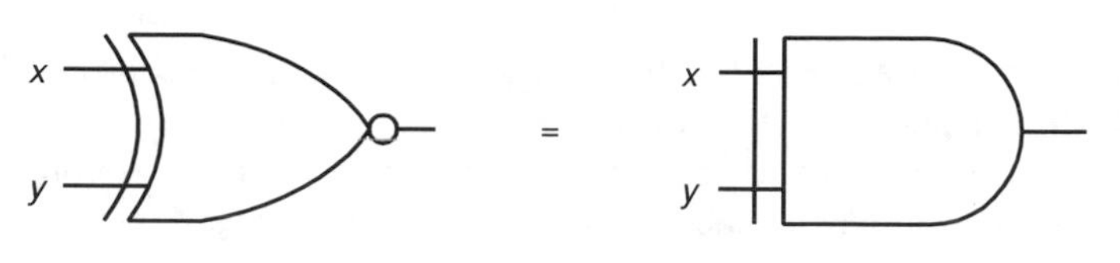

Figure 4.12 NOT XOR is equivalent to COIN.

portrayed not with the circuit symbol shown in Figure 4.10 but as an XOR gate with a NOT bubble on its output as shown on the left-hand side of Figure 4.12.

The circuit diagrams in Figure 4.13 all represent the same function, namely, the coincidence function. In similar fashion the circuit diagrams in Figure 4.14 represent the exclusive OR function.

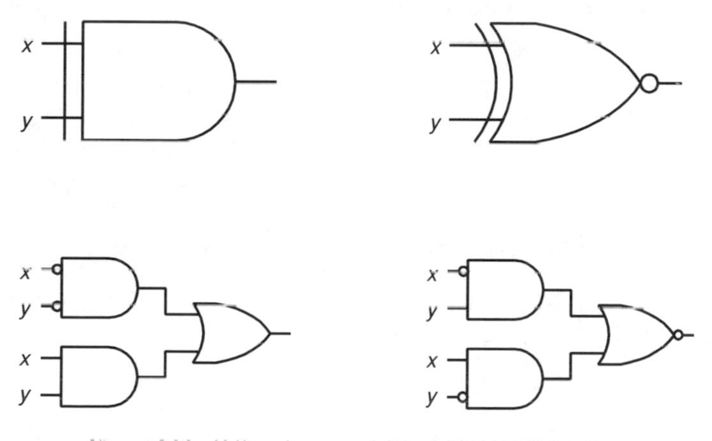

Figure 4.13 Different representations of the COIN function.

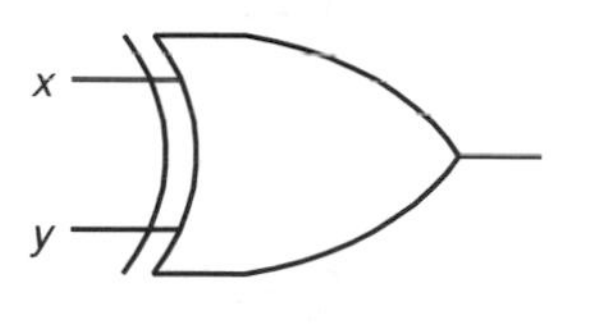

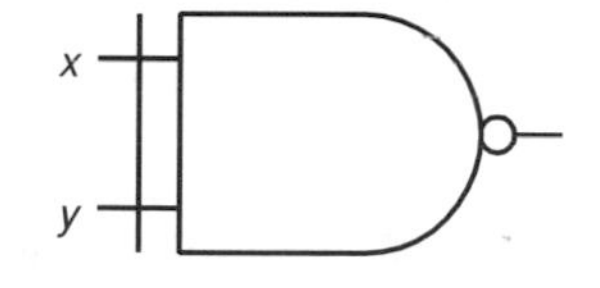

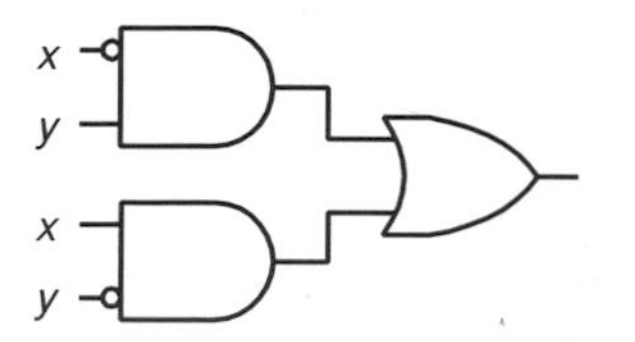

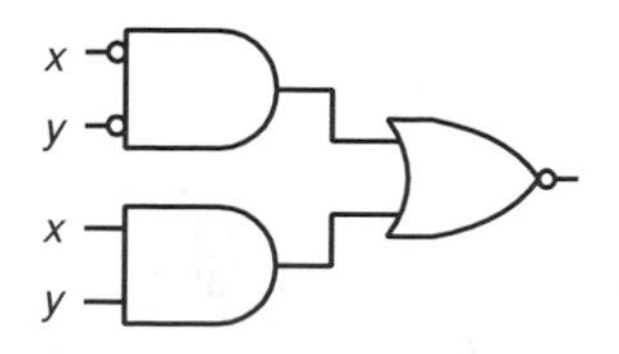

Figure 4.14 Different representations of the XOR function.

Let us look for a moment at the expression $x \odot x$. The coincidence function is 1 if both its inputs are the same, and here both of its inputs are x. x is always equal to itself, whether x is 0 or 1. So $x \odot x$ always equals 1; therefore $x \odot x$ is a tautology (the constant 1 function). Since it only has one variable, x, its truth table (Table 4.12) is only $2^1 = 2$ lines long.

TABLE 4.12

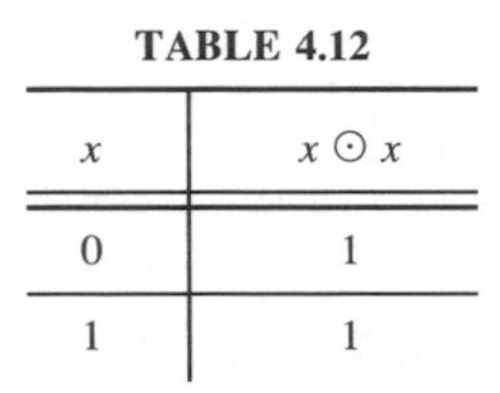

x	$x \odot x$
0	1
1	1

Similarly, $x \oplus x$ is equivalent to the constant 0 function. These situations are illustrated in the circuit diagrams in Figure 4.15.

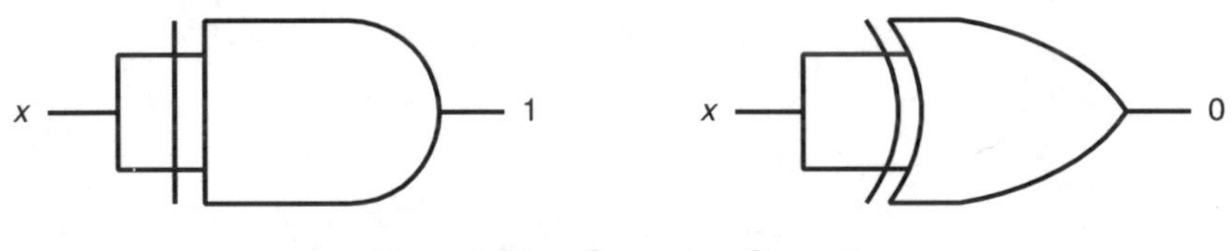

Figure 4.15 $x \odot x = 1; x \oplus x = 0.$

Furthermore, since x and $\bar{x}$ are always different whether x is 0 or 1, $x \odot \bar{x} = 0$, and $x \oplus \bar{x} = 1$ (Fig. 4.16).

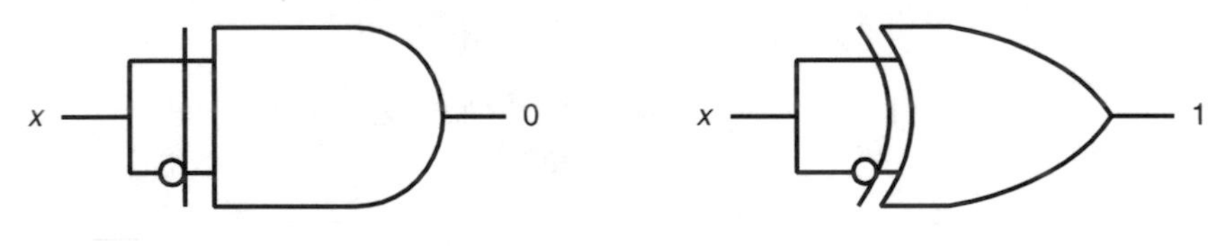

Figure 4.16 $x \odot \bar{x} = 0; x \oplus \bar{x} = 1.$

Knowing that the coincidence function is the complement of the exclusive OR function makes it rather easy to draw the Venn diagram of its corresponding function in the algebra of sets. We simply draw the complement of the Venn diagram we already drew for the exclusive OR function (Fig. 4.17).

4.6 IMPLICATION

The last of the basic two-input Boolean functions we will introduce in this chapter is the **implication** function. Its algebraic symbol is $\rightarrow$. We say that x **implies** y equals 1 if the following condition is true: IF $x = 1$ THEN $y = 1$. Thus the

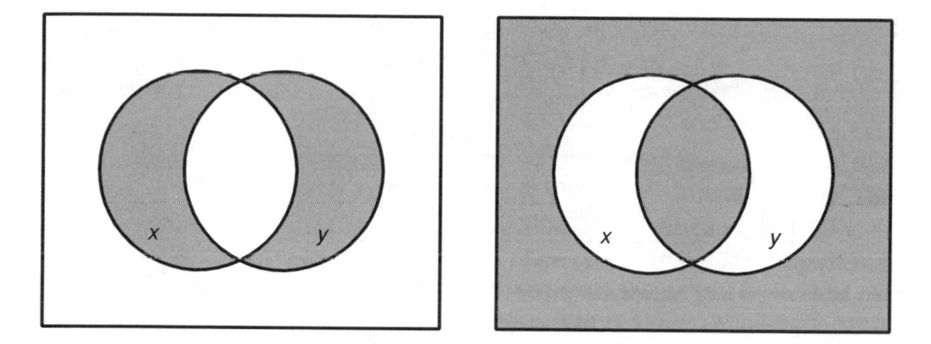

Figure 4.17 Venn diagrams of $(x' \cap y) \cup (x \cap y')$ (exclusive OR) and $(x' \cap y') \cup (x \cap y)$ (coincidence).

function works more or less the same way the English word ''implies'' does. This function is a little different than the functions covered so far, but it has a powerful logical interpretation. A statement of the type ''If . . . then . . .'' is a **conditional statement.**

As an illustration of the workings of the implication function, consider Pavlov's dogs. Ivan Pavlov (1849–1936) was a Russian who worked in the nineteenth century on conditioned psychological and physiological responses, particularly in dogs. He kept some dogs and always rang a bell just before feeding time. Eventually, the dogs got used to this and began to associate the ringing of the bell and the food on an almost reflexive level. As far as the dogs were concerned, the ringing of the bell implied the arrival of food.

To create a truth table of a Boolean function that conforms to this somewhat high-level specification of an implication function, let us use the dogs' assumption that a ringing bell implies the arrival of food and examine the cases in which the assumption is validated as opposed to the cases in which the assumption is disproved. To do this, we create two bits, the first called *bell rings* and the second called *food arrives.* What happens to the dogs' belief that *bell rings* → *food arrives* in each of the four different combinations of these two bits?

If both bits are 1, i.e., the bell rings and food arrives, the dogs' belief is reinforced. The implication holds true. If both bits are off, i.e., no bell rings and no food arrives, the implication still holds true; the dogs hunger happily in the belief that if a bell were to ring, food would surely follow. If *bell rings* = 0 and *food arrives* = 1 (i.e., food arrives even though no bell rang), the implication still holds true, even though this particular situation is not an example of it.

The only situation that should shatter the dogs' belief in the truth of the implication is that in which *bell rings* = 1 and *food arrives* = 0: the bell rings, but no food comes. In this case, and this case only, *bell rings* → *food arrives* = 0, and the implication is rendered false. (As it happens, however, dogs being dogs, Pavlov found that they continued to believe in the truth of the implication even in this situation.) Based on these four cases, we can write the truth table for the implication function as it relates to the bits we discussed as seen in Table 4.13 (the implication function is also that specified by column 11 in Table 4.2). In the expression $x \rightarrow y$, x is called **implicant** and y is called the **consequent.**

TABLE 4.13

Bell Rings	Food Arrives	Bell Rings → Food Arrives
0	0	1
0	1	1
1	0	0
1	1	1

As with AND, OR, and the other Boolean functions discussed so far, according to the principle of assertion, $x \rightarrow y$ is equivalent to the assertion of its own truth, but this assertion may be true or false depending on the particular bits provided as input to the function. The implication function may be thought of as asking ''does x imply y?'' in the same way that AND asks ''Are all inputs 1?'' and OR asks ''Is at least one input 1?'' The output of the function yields the answer according to the given input bits.

The Boolean implication function does differ subtly from our common sense understanding of the word ''implies.'' According to the truth table of the implication function, $0 \rightarrow 0 = 1$. Interpreted using logical propositions, this means that the statement ''If I were made of ice cream then pigs could fly'' is true. Although pigs cannot fly, because I am not and will never be made of ice cream, the question of porcine aerodynamics is moot. Within the realm of Boolean algebra, to say that one thing implies another is not necessarily to suggest that there is any cause and effect relationship between the two; it only means that whenever the implicant is 1, then the consequent is also 1. A by-product of this fact is that anything (even 0) is implied by the constant 0.

However, this property of the implication function is not as counterintuitive as it may first appear. How many times have you heard someone say something like, ''If the Red Sox win the pennant this year, I'm the Dalai Lama''? Assuming that the speaker is not, in fact, the Dalai Lama, the statement is meant to express certainty that the Red Sox will not win the pennant, as only their losing could render the implication true. It should be noted that if the implication function had not been defined to be 1 in the case in which both of its input bits were false, the output of the implication function would simply be an echo of the x input variable.

4.6.1 Arithmetic Interpretation of Implication

Just as we suggested that the AND function could be thought of as the integer minimum function and the OR function could be thought of as the integer maximum function, the implication function has an interpretation in the world of integer arithmetic. It may be thought of as the less-than-or-equal-to function. That is, if we allow ourselves a departure from strict Boolean algebra for a moment and think of 0 and 1 as integers (in which case there is such a thing as "order" and we may say that 0 is less than 1) then we may say that $x \rightarrow y = 1$ only when x is less than or equal to y. A quick look at the truth table confirms that the implication function conforms to this interpretation. In fact, some Boolean algebra texts prefer to denote the implication function by the symbol "$\leq$" rather than the arrow we have chosen to use.

4.6.2 Algebraic Realization of Implication

As we have done with all of the other functions in this chapter, let us realize the implication function in terms of AND, OR, and NOT. According to its truth table, the implication function is 1 in all cases except the case in which $x = 1$ and $y = 0$. So it is 1 whenever $x = 0$, as well as whenever $y = 1$. This suggests that the implication function might be realized by the expression $\bar{x} \vee y$, which is 1 whenever $x = 0$ or $y = 1$. To check the validity of this hypothesis, we will write the truth table for $\bar{x} \vee y$ (Table 4.14).

TABLE 4.14

x	y	$\bar{x}$	$\bar{x} \vee y$
0	0	1	1
0	1	1	1
1	0	0	0
1	1	0	1

The truth tables match, so $\bar{x} \vee y$ is a valid realization of the implication function. The only way $\bar{x} \vee y$ could be 0 is if $\bar{x}$ and y both were 0, or, in other words, if $x = 1$ and $y = 0$.

4.6.3 Circuit Symbol for Implication

Like NAND and NOR, the implication function does not really have its own unique circuit symbol. As the implication function is almost never used by electrical engineers, this is not surprising. We can create a circuit symbol for it using an OR gate and a NOT bubble on one input, thereby creating a circuit diagram for the function $\bar{x} \vee y$ shown in Figure 4.18.

Another way of understanding the implication function is to consider two Boolean functions, a and b, each of which take the same number of inputs. We may say that $a \rightarrow b = 1$ only if there is no row for which there is a 1 in the

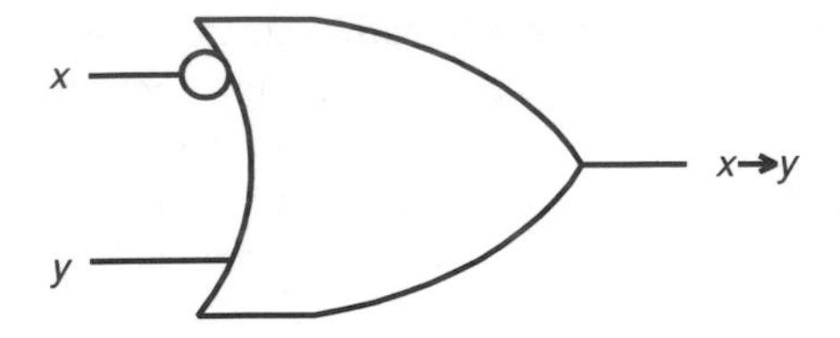

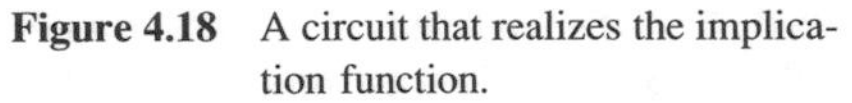
Figure 4.18 A circuit that realizes the implication function.

output column of a's truth table but a 0 in the corresponding row of b's truth table. That is, if $a \rightarrow b = 1$, any time a is 1, b is 1 also. To illustrate, consider the AND and OR functions (Table 4.15).

We may say that $(x \wedge y) \rightarrow (x \vee y) = 1$ because $x \vee y$ is 1 in at least the same rows in the truth table that $x \wedge y$ is 1. The statement, "If $x \wedge y = 1$ then $x \vee y = 1$" is always true, so $(x \wedge y) \rightarrow (x \vee y)$ is a tautology.

However, the reverse is not true. Because there are cases in which $x \vee y = 1$ but $x \wedge y$ does not (e.g., $(x, y) = (0, 1)$), it is not the case that every time $x \vee y = 1, x \wedge y = 1$ also. So $(x \vee y) \rightarrow (x \wedge y)$ does not equal the constant 1 function.

TABLE 4.15

x	y	$x \wedge y$	$x \vee y$
0	0	0	0
0	1	0	1
1	0	0	1
1	1	1	1

4.6.4 Asymmetry of the Implication Function

The implication function is different from all of the other functions that we have looked at so far because it is **asymmetric.** This means that the order of the inputs affects the output. AND, OR, NAND, NOR, XOR, and COIN are all **symmetric;** the order of the input bits makes no difference whatsoever. Thus, for example:

$$x \wedge y = y \wedge x$$

$$x \vee y = y \vee x$$

$$x \uparrow y = y \uparrow x$$

$$x \downarrow y = y \downarrow x$$

$$x \oplus y = y \oplus x$$

$$x \odot y = y \odot x$$

But, $(x \rightarrow y)$ does not equal $(y \rightarrow x)$. They are equal in some cases, but not all. Saying that a bell ringing implies the arrival of food is not the same as saying that the arrival of food implies the ringing of a bell. This is seen clearly in the comparison of the truth tables of $x \rightarrow y$ and $y \rightarrow x$ in Table 4.16.

TABLE 4.16

x	y	$x \rightarrow y$ $(\bar{x} \vee y)$	$y \rightarrow x$ $(\bar{y} \vee x)$
0	0	1	1
0	1	1	0
1	0	0	1
1	1	1	1

Consider the two-input OR function, $x \vee y$. Because whenever either input is 1, then $x \vee y$ is also 1, both inputs individually are implicants of $x \vee y$. Thus $x \rightarrow (x \vee y)$ is a tautology, as is $y \rightarrow (x \vee y)$. Any input of an OR function implies the output of the OR function. In terms of logical propositions, if we already know that Herman is a stockbroker, we know for sure that he is either a stockbroker or a fireman. This situation is illustrated by the circuit diagram in Figure 4.19.

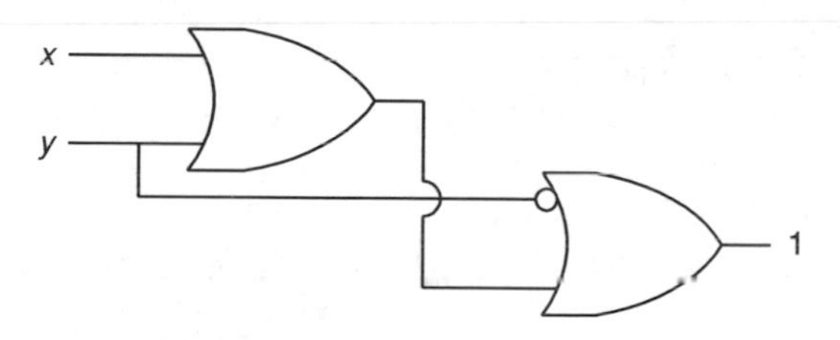

Figure 4.19 y always implies $(x \vee y)$.

Note that the line bearing the output of $x \vee y$ had to "jump" over the y line. This was done so y could be the first input to the implication and $x \vee y$ the second. Otherwise, the output of the implication gate would have been $(x \vee y) \rightarrow y$, instead of $y \rightarrow (x \vee y)$, which, as we saw previously, would have yielded a different result.

In similar fashion, whenever $x \wedge y$ is 1, x must be 1 as well, so $(x \wedge y) \rightarrow x = 1$, as does $(x \wedge y) \rightarrow y$.

4.6.5 Interpreting Implication of Terms of the Algebra of Sets

Given that the implication function is realized by the Boolean algebraic expression $\bar{x} \vee y$, it is easy to draw the Venn diagram of its set algebraic analogue. In fact, it is easier to draw the Venn diagram by breaking it down into separate sets and drawing the sets than it is to "steal" the diagram from the truth table for implication. The set that represents $x' \cup y$ is the union of the set x' and the set y (Fig. 4.20).

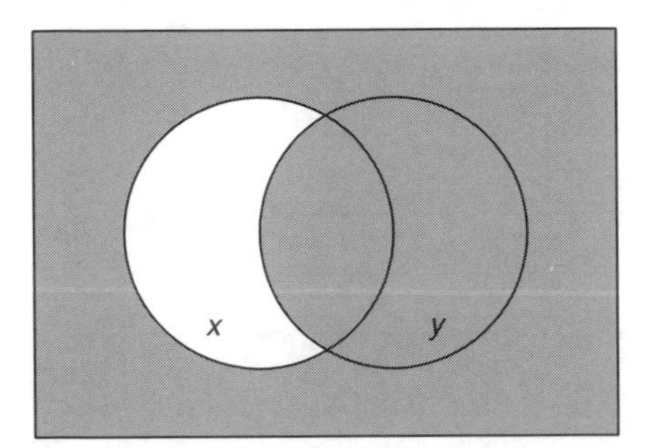

Figure 4.20 $x' \cup y$.

The set algebraic interpretation of the implication function is closely related to the concept of the subset. The set algebraic interpretation of the expression $x \rightarrow y = 1$ is that the set x is a subset of the set y, that is $x \subset y$. For example, to state in terms of the algebra of sets that the set of hammers is a subset of the set of tools is the same as saying in terms of Boolean algebra that if an object is a hammer, that implies that the object is also a tool. We saw that $(x \wedge y) \rightarrow (x \vee y) = 1$. In terms of the algebra of sets, this tells us that the intersection of two sets is always a subset of their union, $(x \cap y) \subset (x \cup y)$.

Similarly, the fact that $x \rightarrow (x \vee y)$ is always true may be interpreted as saying that a set is always a subset of the union of itself and some other set, whatever that other set may be.

In general, to draw the Venn diagram corresponding to a Boolean expression involving implication, we shade in the entire Universe *except* that portion of the implicant's set that is not contained within the consequent's set. The shaded area, then, can be thought of as the maximal Universe in which the implicant set is a subset of the consequent set. Interpreted this way, the Venn diagram of the implication function itself (Fig. 4.20) makes sense beyond its being simply a faithful set algebraic interpretation of the Boolean realization of the implication function, $\bar{x} \vee y$. It is a diagram of the entire Universe except that portion of x that is not a subset of y.

EXERCISE 4.1

A. For each of the following algebraic expressions derive: (a) a circuit diagram, (b) a Venn diagram, (c) a truth table, and (d) an equivalent expression using only AND, OR, and NOT.

1. $(x \uparrow y) \uparrow z$
2. $(x \rightarrow y) \downarrow y$

3. $(x \oplus y) \oplus x$
4. $[(x \downarrow y) \wedge y] \odot (x \downarrow y)$
5. $(x \odot y) \odot (x \odot y)$

B. For each of the following circuit diagrams derive: (a) an algebraic expression, (b) an equivalent algebraic expression using only AND, OR, and NOT, (c) a Venn diagram, (d) a truth table, and (e) a simpler realization, if one exists.

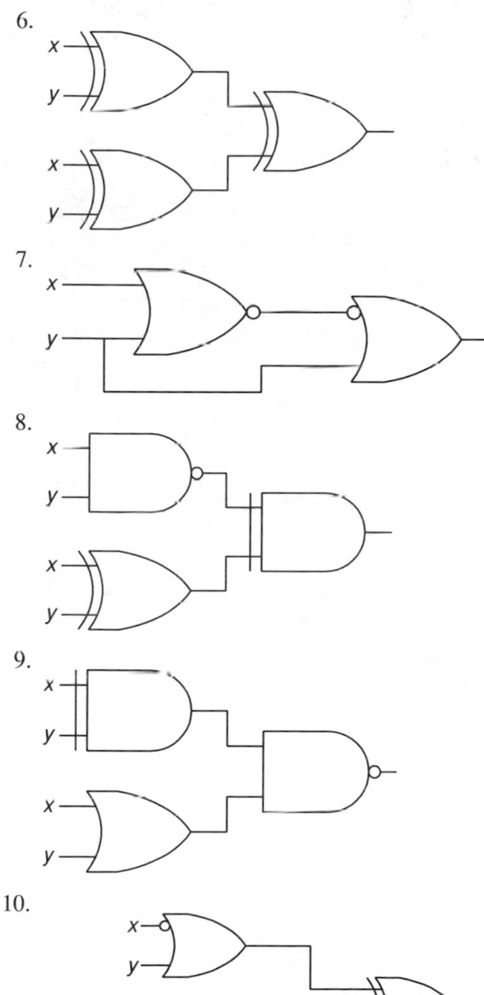

10.

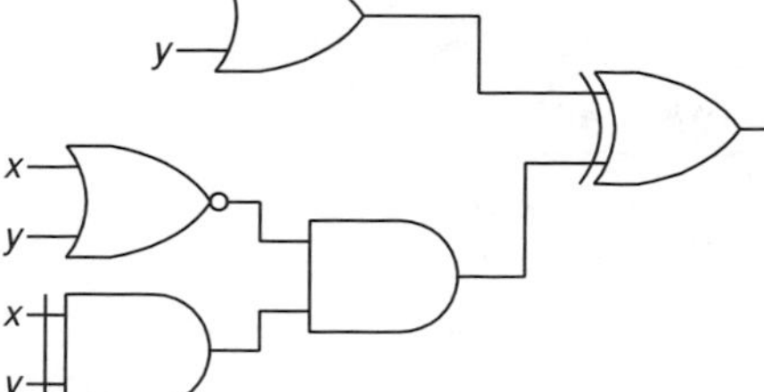

11. Characterize as simply as possible in terms of the functions that you know the functions represented by the following columns in Table 4.2: 10, 12, 5, 3, 4, 13.
12. Expanding on the logic that led us to conclude that there are 16 different two-bit Boolean functions, for any number n, how many n-input Boolean functions are there?
13. Describe in your own words what you need to know about a symmetric Boolean function to characterize it completely (i.e., to write out its entire truth table). For any number n, how many different n-bit symmetric Boolean functions are there?

4.7 OTHER COMPLETE SYSTEMS

We have just considerably expanded the scope of our mathematical system, Boolean algebra, with the addition of NAND, NOR, XOR, COIN, and implication. Moreover, since we could construct each of the new functions out of AND, OR, and NOT, we have preserved the elegance of the system at the same time: we used a relatively small number of primitive functions out of which we were able to build everything else. Anything we construct from now on using any of the new functions could also be built just using AND, OR, and NOT.

4.7.1 XOR, NOT, and 1 as a Complete System

The system of Boolean algebra may be built up from an entirely different base of primitive functions. It turns out that AND, OR, and NOT can themselves be made out of the function XOR and AND and the constant 1 function. To prove this, we only need to show that OR and NOT can be constructed from XOR and AND since AND is already included in XOR, AND, and 1.

First, we can easily make NOT by XORing a bit with 1 as shown in Table 4.17. Since XOR is 0 when its inputs are the same and 1 when they are different, x XOR 1 is 0 if $x = 1$ and 1 if $x = 0$. So x XOR $1 = \bar{x}$. This is illustrated in Figure 4.21.

Now we only need to create OR to show that XOR, AND, and 1 comprise as powerful a mathematical system as AND, OR, and NOT. This is a little more

TABLE 4.17

x	1	$x \oplus 1$
0	1	1
1	1	0

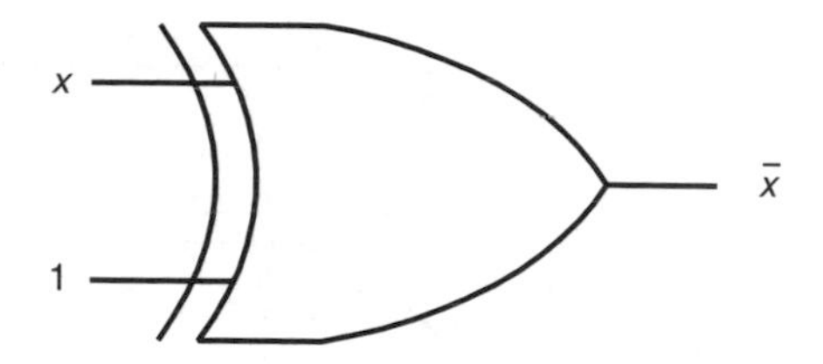

Figure 4.21 $x \oplus 1 = \bar{x}$

complicated. The necessary expression is $((x \wedge y) \oplus y) \oplus x = x \vee y$. To prove this, we need to write the truth table for $[(x \wedge y) \oplus y] \oplus x$ (Table 4.18).

TABLE 4.18

x	y	$x \wedge y$	$(x \wedge y) \oplus y$	$[(x \wedge y) \oplus y] \oplus x$
0	0	0	0	0
0	1	0	1	1
1	0	0	0	1
1	1	1	0	1

The output column is the same as that of OR, thus proving that the OR function can be constructed entirely out of XOR and AND. The circuit diagram illustrating this is shown in Figure 4.22.

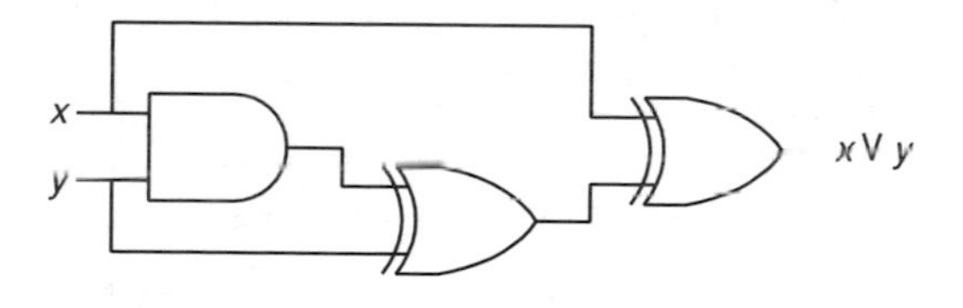

Figure 4.22 $[(x \wedge y) \oplus y] \oplus x = x \vee y$.

4.7.2 NAND as a Complete System

Having shown that AND, OR, and NOT can be constructed using only XOR, AND, and the constant 1, we have proven that if we started with those functions, we could build everything that we could ever build with our AND, OR, and NOT system. We will now go further and show that we can do just as much with the NAND function alone.

Recall the truth table for NAND (Table 4.19). In particular, note that $0 \uparrow 0 = 1$ and $1 \uparrow 1 = 0$. Thus, $x \uparrow x = \bar{x}$, giving us a NOT function constructed using only NAND, as shown in Figure 4.23.

TABLE 4.19

x	y	$x \uparrow y$
0	0	1
0	1	1
1	0	1
1	1	0

Figure 4.23 $x \uparrow x = \bar{x}$.

Since NAND and AND are complementary, all we have to do to make AND is invert NAND, which we can do using the NOT we just built (Fig. 4.24):

$$x \wedge y = \overline{(x \uparrow y)} = (x \uparrow y) \uparrow (x \uparrow y)$$

Finally, we need to create OR out of NANDs. OR is always 1 except when both inputs are 0; NAND is always 1 except when both inputs are 1. Thus, if we invert the inputs of NAND, a pair of 0's will appear to the NAND as a pair of 1's and yield a 0 accordingly (Table 4.20).

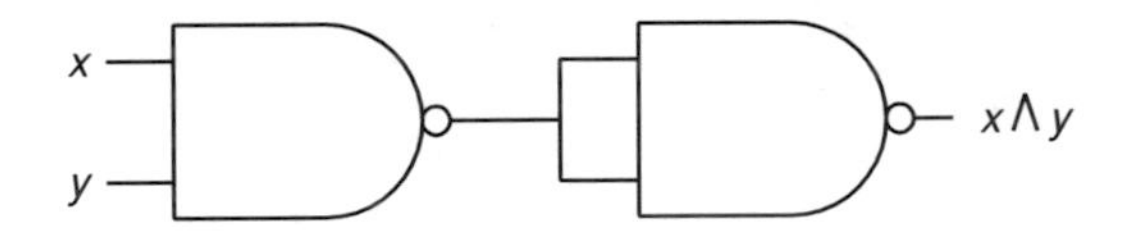

Figure 4.24 $(x \uparrow y) \uparrow (x \uparrow y) = x \wedge y$.

TABLE 4.20

x	y	$\bar{x}$	$\bar{y}$	$\bar{x} \uparrow \bar{y}$
0	0	1	1	0
0	1	1	0	1
1	0	0	1	1
1	1	0	0	1

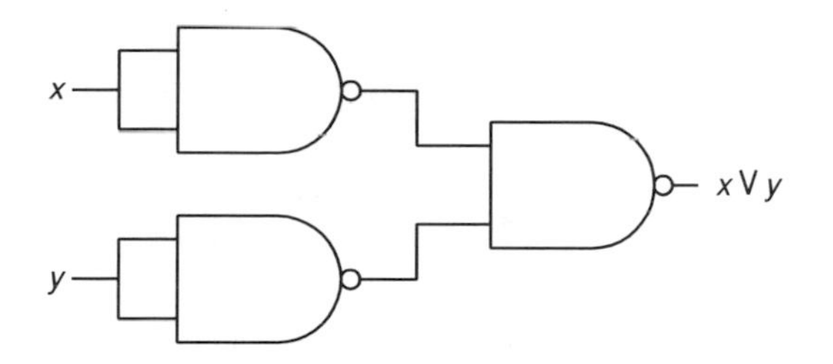

Figure 4.25 $(x \uparrow x) \uparrow (y \uparrow y) = x \vee y$.

As the truth table shows, $\bar{x}$ NAND $\bar{y}$ is the same function as x OR y. From here, the rest is easy. We know how to invert x and y using NAND (Fig. 4.25):

$$x \vee y = \bar{x} \uparrow \bar{y} = (x \uparrow x) \uparrow (y \uparrow y).$$

We have shown that we can make AND, OR, and NOT out of NAND. Furthermore, we can make all the other Boolean functions that we know about as well, since they in turn could be made with AND, OR, and NOT. In addition to NAND and XOR, AND, and 1, there are other sets of primitive functions out of which we could build AND, OR, and NOT and thus use as a basis for our system of Boolean algebra. Some of the resulting systems, like the NAND system, would begin with fewer primitive functions than does the AND, OR, and NOT system and therefore, would, be more technically elegant. Nonetheless, we shall continue to use AND, OR, and NOT because they constitute a more intuitively clear logical basis for Boolean algebra and are easier to work with. Consider, for example, the prospect of making COIN out of NAND. It can be done, as we have proven previously, but the resulting function is dauntingly unwieldy.

EXERCISE 4.2

1. Realize NAND, NOR, COIN, and implication in terms of XOR, AND, and constant 1 only.
2. Realize NOR, XOR, COIN, and implication in terms of NAND only.

   For each of the following algebraic expressions derive: (a) a circuit diagram, (b) a Venn diagram, and (c) a truth table. Then (d) convert the expression to an equivalent expression using only XOR and AND, and (e) derive a circuit diagram for it.

3. $(x \rightarrow y) \downarrow z$
4. $(x \odot y) \vee (x \uparrow y)$
5. $(x \downarrow y) \vee (x \vee y)$
6. $(x \vee y) \odot (x \uparrow y)$
7. $\overline{(x \odot y)}$

For each of the following algebraic expressions: (a) write the truth table and b) convert the expression to an equivalent expression using only NAND.

8. $(x \oplus y) \downarrow z$

9. $\bar{x} \vee \bar{y}$

10. $x \rightarrow \bar{y}$

11. $(x \rightarrow y) \wedge (x \vee y)$

12. Prove that NOR itself is a complete system, i.e., build AND, OR, and NOT using NOR alone.

5

Realizing Any Boolean Function with AND, OR, and NOT

In the previous chapters we have seen several different ways of representing and thinking about Boolean functions. We can write the truth table for any function, whether it is presented in the form of a Boolean algebraic expression, a set algebraic expression, a Venn diagram, or a circuit diagram. In this chapter we will work in the other direction, discovering ways to derive a Boolean function for any truth table.

This is a powerful skill. It allows us to custom design a function (or a digital circuit) by first specifying its output in the form of a truth table, then putting together the correct combinations of AND, OR, and NOT to realize the desired function. We had a taste of this in the chapter 4 when we found the algebraic expressions for the XOR, COIN, and implication functions in terms of AND, OR, and NOT, just by looking at their truth tables. We will now generalize and formalize the methods we used then.

5.1 MINTERMS

First we will consider a special class of Boolean functions: those for which one and only one combination of input bits yields a 1, and all the other combinations yield a 0. AND is an example of such a function. Another example is represented in the three-input truth table shown in Table 5.1.

There is only one 1 in the output column of this function's truth table. The function is 1 only when $(x, y, z) = (1, 0, 1)$ and all other combinations of inputs

TABLE 5.1

x	y	z	
0	0	0	0
0	0	1	0
0	1	0	0
0	1	1	0
1	0	0	0
1	0	1	1
1	1	0	0
1	1	1	0

cause this function to yield a 0 as output. Put another way, this function will be 1 only when $x = 1$ AND $y = 0$ AND $z = 1$, and it will be 0 at all other times.

Translated into the terms of Boolean algebra by way of the principle of assertion, the function is 1 when x AND (NOT y) AND z. Thus $x \wedge \bar{y} \wedge z$ is the algebraic expression of the function represented by the truth table shown in Table 5.1. To verify this, we will write the truth table for $x \wedge \bar{y} \wedge z$ as shown in Table 5.2. As predicted, the output column of $x \wedge \bar{y} \wedge z$ matches the output column of our mystery function perfectly: they are the same function.

In general then, to find the algebraic expression for any truth table that only has one 1 in its entire output column:

1. Write out the function's input variables with ANDs between them, e.g., $x \wedge y \wedge z$.
2. Draw NOT bars over those variables that are 0 in that combination of input bits that makes the function 1.

TABLE 5.2

x	y	z	$\bar{y}$	$x \wedge \bar{y} \wedge z$
0	0	0	1	0
0	0	1	1	0
0	1	0	0	0
0	1	1	0	0
1	0	0	1	0
1	0	1	1	1
1	1	0	0	0
1	1	1	0	0

In the function above, the one combination of inputs that made the function 1 was $(x, y, z) = (1, 0, 1)$. Since the middle variable, y, is 0 in this combination, we drew a NOT bar over the y in our expression, yielding $x \wedge \bar{y} \wedge z$.

This method can be applied no matter how many input bits there are, as long as there is only one combination of those input bits that yields a 1 as output. For example, imagine that we have a truth table that represents a six-input Boolean function. The truth table would be $2^6 = 64$ lines long. (See Appendix B for a chart of the powers of 2.) Now imagine that the output column for 63 of the rows contains a 0, and there is a 1 in the output column only for the row in which the inputs are $(a, b, c, d, e, f) = (1, 1, 0, 1, 0, 0)$ (Table 5.3).

TABLE 5.3

a	b	c	d	e	f	
0	0	0	0	0	0	0
0	0	0	0	0	1	0
0	0	0	0	1	0	0
1	1	0	0	1	0	0
1	1	0	0	1	1	0
1	1	0	1	0	0	1
1	1	0	1	0	1	0
1	1	0	1	1	0	0
1	1	1	1	1	0	0
1	1	1	1	1	1	0

The whole 64-line truth table would be too long to write in its entirety, so it has been abbreviated here. The dots represent the parts left out and all rows not shown contain 0's in their output columns. To find the algebraic expression of the function represented by this truth table, first write down all the input variables with ANDs between them: $a \wedge b \wedge c \wedge d \wedge e \wedge f$. Then draw NOT bars over those variables that are 0 in the one combination of inputs that makes the function 1. For this function, that combination is $(a, b, c, d, e, f) = (1, 1, 0, 1, 0, 0)$ in which the variables c, e, and f are 0. Thus, we draw NOT bars over the c, e, and f in our expression, and we are done. The algebraic realization of the desired function is $a \wedge b \wedge \bar{c} \wedge d \wedge \bar{e} \wedge \bar{f}$. This function is 0 for all combinations of input bits 000000 to 111111, except that it is 1 for the single combination $(a, b, c, d, e, f) = (1, 1, 0, 1, 0, 0)$.

Such a function is called a **minterm** for *minimum term* because the function is almost always 0; it is 1 for only one particular combination of its n input bits out of 2^n possible combinations. A minterm consists of all input variables of a given function, complemented or not, ANDed together.

Sometimes we will use a shorthand notation whereby a function that is a minterm is named for the base-10 representation of the particular combination of binary inputs that make the minterm 1. This number will also, not coincidentally, be the number that designates the only row of the truth table in which a 1 appears in the output column. The particular minterm in the example above would be called minterm 52 (written M_{52}) because the 1 appears in the output column of row 52 of the truth table; the particular combination of input bits that make the function 1, taken together form the number 110100_2, which is equal to 50_{10}. Thus the expression M_{52} means $a \wedge b \wedge \bar{c} \wedge d \wedge \bar{e} \wedge \bar{f}$ and both are ways of representing the Boolean function that will be 1 when and only when its inputs bits are $(a, b, c, d, e, f) = (1, 1, 0, 1, 0, 0)$.

5.1.1 Decoder Example

A decoder is a piece of electronic circuitry that can be made out of AND, OR, and NOT gates. For a given number n (which varies depending on the type of decoder) a decoder has n input lines and 2^n output lines. Thus, if a particular decoder circuit has three input lines, it has $2^3 = 8$ output lines. Each input line specifies one bit of an number in base 2, and the decoder simply sends out a 1 on the output line corresponding to the n-bit input number taken as a whole. All other outputs are 0.

For example, let us call the inputs to a three-input, eight-output decoder (called a 3-to-8 decoder) x, y, and z, and let us consider x to be the most significant bit of the binary input number and z to be the least significant bit. If this decoder were given the input combination $(x, y, z) = (1, 0, 1)$, it would send a 1 out on output line five ($101_2 = 5_{10}$), and 0's out on all other output lines (Fig. 5.1).

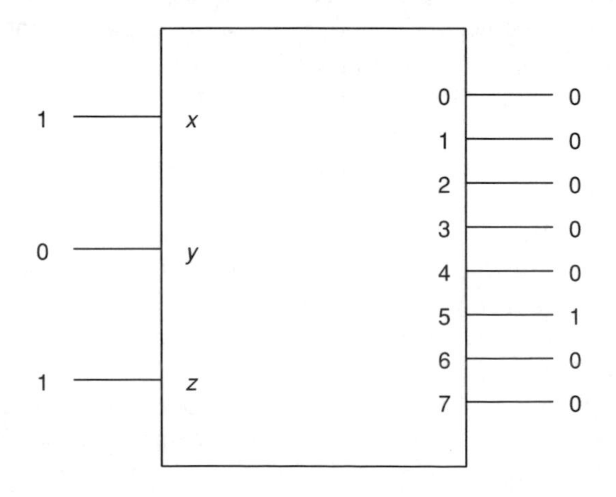

Figure 5.1 3-to-8 decoder given (1, 0, 1) as input.

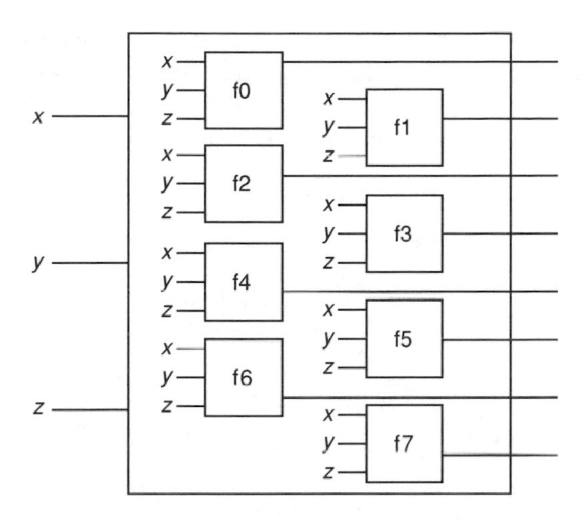

Figure 5.2 3-to-8 decoder showing separate functions for all output lines.

Only one output line is ever 1 at a time, that which corresponds to the input number.

How would we build a 3-to-8 decoder out of AND, OR, and NOT gates? Since there are eight output lines, we really need eight separate functions, one for each output line. All of these functions will use the same three inputs, x, y, and z. Let us call the eight functions f0 through f7, so for example, the output line of function f3 will be output line three of the decoder; its inputs will be x, y, and z. We want the output of f3 to be 1 only when $(x, y, z) = (0, 1, 1)$ 3_{10}, and 0 at all other times. Figure 5.2 shows the 3-to-8 decoder so far, showing the individual functions f0–f7. The truth table for each of the eight functions is shown in Table 5.4.

TABLE 5.4

x y z	f0	f1	f2	f3	f4	f5	f6	f7
0 0 0	1	0	0	0	0	0	0	0
0 0 1	0	1	0	0	0	0	0	0
0 1 0	0	0	1	0	0	0	0	0
0 1 1	0	0	0	1	0	0	0	0
1 0 0	0	0	0	0	1	0	0	0
1 1 0	0	0	0	0	0	0	1	0
1 1 1	0	0	0	0	0	0	0	1

The output columns of all eight functions together form a solid block of 0's with a diagonal stripe of 1's. Since each function is 1 for one and only one combination of inputs, each can be realized by a minterm. In fact, the functions f0–f7 are realized by the minterms M_0–M_7 respectively:

$$f0 = M_0 = \bar{x} \wedge \bar{y} \wedge \bar{z},$$

$$f1 = M_1 = \bar{x} \wedge \bar{y} \wedge z,$$

$$f2 = M_2 = \bar{x} \wedge y \wedge \bar{z},$$

$$f3 = M_3 = \bar{x} \wedge y \wedge z,$$

$$f4 = M_4 = x \wedge \bar{y} \wedge \bar{z},$$

$$f5 = M_5 = x \wedge \bar{y} \wedge z,$$

$$f6 = M_6 = x \wedge y \wedge \bar{z}$$

$$f7 = M_7 = x \wedge y \wedge z.$$

Finally, Figure 5.3 shows the circuit diagram of a 3-to-8 decoder (the actual connections between the inputs *x, y,* and *z* into the whole decoder and the inputs *x, y,* and *z* into each of the individual eight functions inside are not shown because the resulting tangle of split inputs jumping over each other would render the diagram difficult to understand).

Electrical engineers often find decoders useful, for example when they must design a piece of circuitry that will trigger certain events on the basis of specific values of a binary number provided by another piece of circuitry (e.g., digital

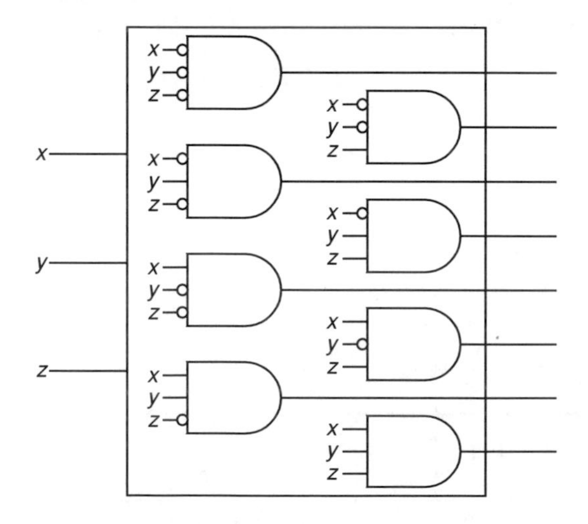

Figure 5.3 3-to-8 decoder showing realizations of functions for all output lines.

thermometer or timer). The binary number is fed into the input of an appropriately sized decoder, and the outputs corresponding to the desired values of the input number are used as electronic triggers.

Consider the two-input bit minterm M_0. Its output is 1 only when its input bits are $(x, y) = (0, 0)$, so its algebraic expression is $\bar{x} \wedge \bar{y}$. Its truth table is shown in Table 5.5. But isn't this just the familiar NOR function, whose algebraic expression is $\overline{(x \vee y)}$? Clearly, $\overline{(x \vee y)}$ and $\bar{x} \wedge \bar{y}$ are two equivalent representations of the same function. Such equivalences will be explored more fully in later chapters.

TABLE 5.5

x	y	
0	0	1
0	1	0
1	0	0
1	1	0

5.2 REALIZING ANY BOOLEAN FUNCTION USING MINTERMS

While our knowledge of minterms only allows us to realize Boolean functions that yield a 1 for just one possible combination of inputs, we may now apply this knowledge to derive a realization of any Boolean function from its truth table. This involves writing down the minterms of each of the combinations of inputs that make the function 1, then putting ORs between the resulting minterms.

This was essentially the method we used when we introduced the XOR function and showed how to realize it in terms of AND, OR, and NOT. Now that we have an explicit method, however, we can derive the realization of XOR more formally. We begin with its truth table (Table 5.6).

TABLE 5.6

x	y	$x \oplus y$
0	0	0
0	1	1
1	0	1
1	1	0

There are two rows in the truth table in which there is a 1 in the output columns, rows 1 and 2, in which $(x, y) = (0, 1)$ and $(x, y) = (1, 0)$. The minterm for the first of these is M_1, or $\bar{x} \wedge y$, and the minterm for the second is M_2, or $x \wedge \bar{y}$. The final realization of the XOR function is derived by writing both of these minterms with OR between them: $M_1 \vee M_2 = (\bar{x} \wedge y) \vee (x \wedge \bar{y})$. Indeed, this is the same realization that we arrived at in chapter 4.

Suppose you are serving ice cream at your five-year-old nephew's birthday party. You have a half gallon of ice cream that is divided into three sections: one each of chocolate, strawberry, and vanilla. Louie, one of your nephew's friends, wants some ice cream, but he claims to be allergic to strawberry. He will, however, happily take either chocolate or vanilla (or both). We will draw a truth table that represents the different ways we could feed Louie ice cream that would make him happy.

First, we know that we have three flavors of ice cream that conceivably could be present or absent in Louie's bowl. Assign one bit to each flavor, and for each flavor bit let a 0 mean that we will not serve Louie that flavor and a 1 mean that we will serve him that flavor. We will call the bits c, s, and v and arbitrarily assign the following order to the bits: $[c, s, v]$.

Since there are three input bits, there are $2^3 = 8$ ways we could possibly feed Louie ice cream, ranging from $(c, s, v) = (0, 0, 0)$ (no flavors at all, or an empty bowl for Louie) to $(c, s, v) = (1, 1, 1)$ (we give him some of each flavor). Here then is our truth table, with 1's in the output columns of those rows that represent combinations of flavors that would make Louie happy and 0's in those rows that would not Table 5.7.

There are 1's in the output columns only in those rows in which the middle bit (strawberry) is 0. The output bit in the first row (no ice cream at all) is also 0 because Louie definitely wants some ice cream. This leaves only the rows in

TABLE 5.7

c	s	v	Louie Happy?
0	0	0	0
0	0	1	1
0	1	0	0
0	1	1	0
1	0	0	1
1	0	1	1
1	1	0	0
1	1	1	0

which $(c, s, v) = (0, 0, 1), (1, 0, 0)$, and $(1, 0, 1)$—the rows that represent just vanilla, just chocolate, and chocolate and vanilla together—the only combinations that will satisfy young Louie. Now let us find the algebraic expression that represents the function specified by Louie's truth table. To do this, we write out the minterms for the three rows with 1's in their output columns with ORs between them:

$$M_1 \vee M_4 \vee M_5 = (\bar{c} \wedge \bar{s} \wedge v) \vee (c \wedge \bar{s} \wedge \bar{v}) \vee (c \wedge \bar{s} \wedge v).$$

The circuit diagram for this function is shown in Figure 5.4.

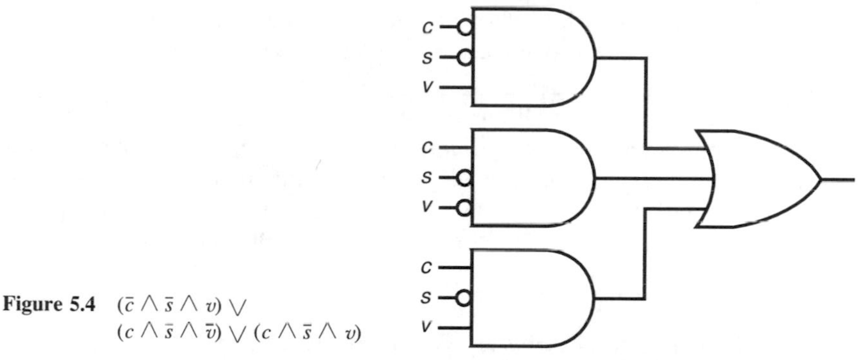

Figure 5.4 $(\bar{c} \wedge \bar{s} \wedge v) \vee (c \wedge \bar{s} \wedge \bar{v}) \vee (c \wedge \bar{s} \wedge v)$

This function, when supplied with different combinations of the chocolate, strawberry and vanilla bits will yield a 0 or a 1, depending on whether the combinations of inputs will satisfy Louie's palate.

5.3 SUM-OF-PRODUCTS EXPRESSIONS

We call a realization of a Boolean function made of minterms ORed together a **minterm realization** or **minterm expression** for that function. All minterm realizations have the same basic form. To use the analogy between the Boolean functions of AND and OR and the arithmetic functions of multiplication and addition, all minterm expressions consist of one or more products added (ORed) together. Each product in turn consists of one or more variables (complemented or uncomplemented) ANDed together. All Boolean expressions that consist of such products added together are called, appropriately enough, **sum-of-products (SOP)** expressions. Sometimes a product in an SOP expression will contain only one variable, in which case the AND is omitted. Similarly, an SOP expression may consist of only one product, in which case the OR is omitted. To carry this to a ridiculous extreme, the expression x itself could be considered an SOP expression of only one product, which itself consists of only one variable. In an SOP expression, individual variables may be complemented, but no larger units may be. That is, $(a \wedge \bar{b} \wedge \bar{c}) \vee (a \wedge b \wedge \bar{c}) \vee (\bar{a} \wedge b \wedge c)$ is an SOP expression, but $(a \wedge \bar{b} \wedge \bar{c}) \vee \overline{(a \wedge b \wedge \bar{c})} \vee (\bar{a} \wedge b \wedge c)$ is not.

Minterm expressions are SOP expressions, but not all SOP expressions are minterm expressions. In a minterm expression each product contains every one of the variables present in the entire expression, but this is not necessarily the case for all SOP expressions. Thus $(\bar{c} \wedge \bar{s} \wedge v) \vee (c \wedge \bar{s} \wedge \bar{v}) \vee (c \wedge \bar{s} \wedge v)$ is a minterm expression, but $(a \wedge \bar{d}) \vee (a \wedge b \wedge \bar{c} \wedge d) \vee c$ is not, although both are SOP expressions. The second expression is not a minterm expression because while the variables *a, b, c,* and *d* appear in the expression, only one of the three products contains all four of these variables, namely $(a \wedge b \wedge \bar{c} \wedge d)$. It is worth noting that an individual product in an SOP expression *implies* the expression of which it is a component, as any input to an OR implies the output of the OR.

The circuit diagrams of all minterm expressions have a certain uniformity of appearance. They contain one or more AND gates (never more than 2^n for a function with n input bits), all of which take all of the inputs to the function as their inputs (although some inputs may be negated). The outputs of all of these AND gates are used as the inputs of a single OR gate, the output of which is the output of the entire function.

5.3.1 Realization of Any Boolean Function Using a Decoder

We can also realize any n-bit Boolean function using an n-to-2^n decoder and an OR gate. If we wish to realize an n-bit Boolean function for which we know the truth table, we can feed the input bits into an n-to-2^n decoder and OR together the output lines from the decoder that correspond to the rows in the truth table that have 1's in their output columns. The output from the OR is the output of the desired function. For example, we can realize the function that describes Louie's taste in ice cream (specified by the truth table listed in Table 5.7) as shown in Figure 5.5.

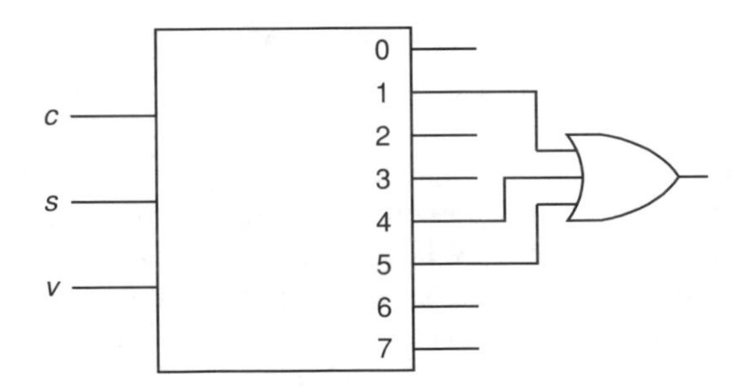

Figure 5.5 Decoder-based realization of $(\bar{c} \wedge \bar{s} \wedge \bar{v}) \vee (\bar{c} \wedge \bar{s} \wedge \bar{v}) \vee (\bar{c} \wedge \bar{s} \wedge v)$.

We implemented the decoder by realizing a function corresponding to each output line with a different minterm. Since each output line from the decoder represents one minterm, it should be clear that the decoder-based realization of any function is just another way of writing the minterm realization of that function, rather than being an entirely different realization. The difference between them is one of representation, as seen in the "transparent" diagram of the decoder realization shown in Figure 5.6.

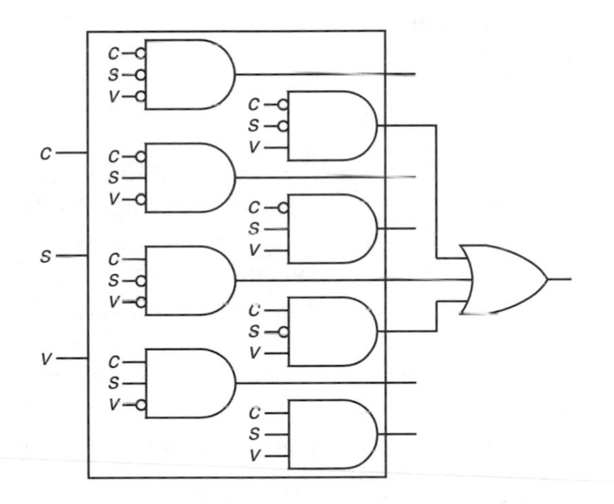

Figure 5.6 "Transparent" view of decoder-based realization of $(\bar{c} \wedge \bar{s} \wedge v) \vee (c \wedge \bar{s} \wedge \bar{v}) \vee (c \wedge \bar{s} \wedge v)$.

5.4 THE SEVEN-SEGMENT DISPLAY

Computers and other digital circuits deal with numbers in base 2. Any number added, subtracted, multiplied, or in any way manipulated or stored by a computer consists solely of zeros and ones. For humans to be able to read numbers produced by computers easily, there must be some sort of human-readable interface between the part of the machine that does the work and the human who benefits from that work. The seven-segment display is a simple example of such a user interface.

If you look at the face of a digital watch, you will see that each digit is actually made up of horizonal and vertical bars in the form of the numeral 8. Each digit, from 0 to 9 is formed by selectively darkening (turning on) particular bars and leaving others transparent (turning them off). In the case of the numeral 8 itself, all of the bars are turned on (Fig. 5.7).

When your digital watch says 12:16, somewhere inside its memory are four numbers: 0001, 0010, 0001, 0110—the binary representations of 1,2,1, and 6,

Figure 5.7 Seven-segment display.

respectively. There is also circuitry in your watch that knows how to take these base-2 numbers as input and turn on or off the correct bars on the watch face to make the digits "12:16" show up so that you can read them. It does this by sending a 1 to the bars that should be on and a 0 to the ones that should be off. The bars are made of a liquid crystal that will darken upon receipt of an electrical "1" and become transparent upon receipt of an electrical "0". An explanation of how the electricity causes the bars to darken lies deep within the realm of chemical and electrical engineering and is beyond the scope of this book. We are, however, interested in the circuitry that figures out when to send a 1 and when to send a 0 to each of the bars.

Consider the rightmost digit in our example, the one that is a six in the time reading 12:16. Associated with this digit on the watch face is circuitry that is currently receiving the bits 0110 (6_{10}) from the timekeeping circuitry (which is also beyond the scope of this book). This display circuitry has one output line connected to each of the seven bars in the rightmost digit. There is a separate Boolean function for each of these seven output lines, all of which take the same four bits as input (as of this moment, 0110). There is identical circuitry for the other three digits on the watch face as well.

Since we can use minterms to realize any Boolean function that we can write a truth table for, to realize each of the seven functions required to display a digit all we need is truth tables describing their operations. The first step toward this goal is to determine exactly how many input bits the function must be able to handle. This will tell us how long the truth table will be. Since the watch handles base-10 digits, we know that each digit on the display must be capable of displaying any digit between 0 and 9 inclusive (i.e., including 0 and 9 themselves). We may assume then that the timekeeping circuitry inside the watch will never send us a number greater than nine. How many bits are needed to ensure that we can represent any number from 0–9?

Two bits allow us to represent the numbers 00_2–11_2 (0–3), which is clearly inadequate. Three bits give us 000_2–111_2 (0–7), just two shy of what we need. Four bits, however, cover our needs and then some. With four bits, we can represent any number in the range 0000_2–1111_2 (0–15). Since our circuitry will never have to deal with a number greater than nine, by using four bits, in a sense, our circuitry is wasting some possible values. Nevertheless, four bits is the minimum number of bits that will be able to deliver the numbers zero through nine to our display.

Now we have four bits from the timekeeping circuitry and we want seven functions, all of which take these same four bits as input. Each of these functions is responsible for darkening (or not darkening) one of the seven individual bars on a single digit on the watch display. We will call the four input bits *a, b, c,* and *d,* and the seven functions f0–f6. The situation so far is shown in Figure 5.8.

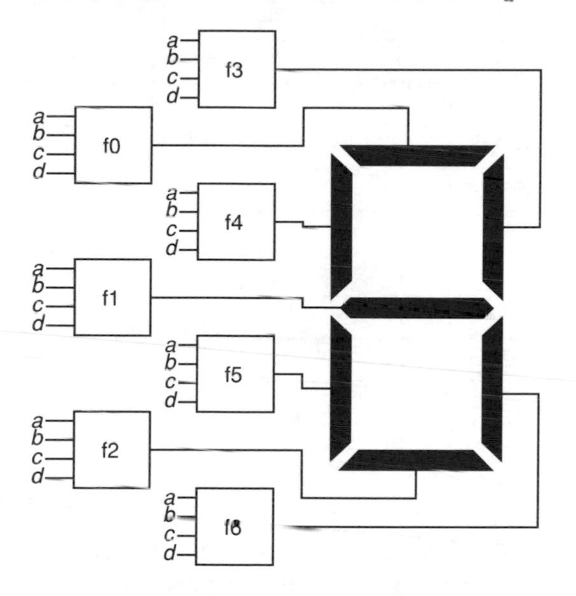

Figure 5.8 Seven-segment display with each controlling function.

Our job now is to create truth tables for the functions f0–f6, and then find minterm expressions that realize these functions in terms of AND, OR, and NOT. As an example, let us consider f5, the function that controls the lower lefthand vertical bar (The other six functions are left as exercises at the end of the section). The function f5 takes four bits as input, representing the numbers zero (0000) through nine (1001). When should it be 1? To help us find out, we will draw the truth table and number each row with a seven-segment display representation of that row's number in base 10 (Table 5.8).

Since this function will never be given numbers greater than nine (1001) as input, its output for the inputs 10 (1010)–15 (1111) is of no concern to us. For this reason, the output of the function in these cases is called a **don't care** and a "d" is put in the output column as a place holder instead of a 0 or a 1.

To find the proper output bit for each of the first ten rows, we look at the

TABLE 5.8

	a	*b*	*c*	*d*	f5
0	0	0	0	0	
1	0	0	0	1	
2	0	0	1	0	
3	0	0	1	1	
4	0	1	0	0	
5	0	1	0	0	
6	0	1	1	0	
7	0	1	1	1	
8	1	0	0	0	
9	1	0	0	1	
–	1	0	1	0	
–	1	0	1	1	
–	1	1	0	0	
–	1	1	0	1	
–	1	1	1	0	
–	1	1	1	1	

digit at the left of the row in the truth table. The numeral 3 should not have the lower lefthand vertical bar darkened, so the function controlling that bar, f5, must produce a 0 when given the input 0011. Accordingly, we write a 0 in the output column of row three in the truth table. In similar fashion we can determine the output bit of f5 for each row, 0–9 (Table 5.9).

We have our function fully specified, except for the don't cares, which we don't care about. It does not matter what our function produces as output when given these combinations of input bits because it will never be handed those particular combinations by the timekeeping circuitry. It is now a simple matter to create a minterm realization of f5. Looking at the truth table, we see that there are four 1's in rows 0, 2, 6, and 8. In all other rows, the output bit is either 0 or *d.* Thus the function f5 is realized by $M_0 \vee M_2 \vee M_6 \vee M_8$ (AND symbols are omitted for clarity):

$$\bar{a}\bar{b}\bar{c}\bar{d} \vee \bar{a}\bar{b}c\bar{d} \vee \bar{a}bc\bar{d} \vee a\bar{b}\bar{c}\bar{d}.$$

We are given the liberty of making don't care outputs 0 and 1 depending

TABLE 5.9

	a	b	c	d	f5
0	0	0	0	0	1
1	0	0	0	1	0
2	0	0	1	0	1
3	0	0	1	1	0
4	0	1	0	0	0
5	0	1	0	1	0
6	0	1	1	0	1
7	0	1	1	1	0
8	1	0	0	0	1
9	1	0	0	1	0
–	1	0	1	0	d
–	1	0	1	1	d
–	1	1	0	0	d
–	1	1	0	1	d
–	1	1	1	0	d
–	1	1	1	1	d

on which choice makes it easier for us to realizc the function. Because we are using minterm expressions, it is easier if we let the don't care outputs all be 0.

Although our method of minterm realization of Boolean functions is powerful, it is not perfect. Considcr the minterm expression of the function whose truth table appears in Table 5.10. This function is realized by the minterm expression $(\bar{x} \wedge y) \vee (x \wedge \bar{y}) \vee (x \wedge y)$. However, this is simply the OR function, much

TABLE 5.10

x	y	
0	0	0
0	1	1
1	0	1
1	1	1

more compactly realized by the expression $x \vee y$. Clearly, minterms are not the most efficient way to realize all Boolean functions.

EXERCISE 5.1

A. Write (a) the minterm realization, (b) the Venn diagram, and (c) the circuit diagram of the functions specified by each of the following truth tables.

1.

x	y	
0	0	1
0	1	1
1	0	0
1	1	1

2. (Remember the trick to drawing four-set Venn diagrams.)

a	b	c	d	
0	0	0	0	0
0	0	0	1	1
0	0	1	0	1
0	0	1	1	0
0	1	0	0	1
0	1	0	1	1
0	1	1	0	0
0	1	1	1	0
1	0	0	0	0
1	0	0	1	1
1	0	1	0	0
1	0	1	1	1
1	1	0	0	1
1	1	0	1	1
1	1	1	0	0
1	1	1	1	0

3.

x	y	z	
0	0	0	0
0	0	1	1
0	1	0	1
0	1	1	0
1	0	0	1
1	0	1	0
1	1	0	0
1	1	1	0

4.

x	y	z	
0	0	0	1
0	0	1	1
0	1	0	1
0	1	1	1
1	0	0	1
1	0	1	1
1	1	0	0
1	1	1	1

5. Write the minterm realizations of the remaining six functions needed to implement a seven-segment display.

5.5 MAXTERMS

In this section we will deal with **maxterms** which are functions that yield 0 for only one combination of input bits and 1 for all other combinations. Just as we ORed minterms together to realize Boolean functions in the last section, we will see that we can AND maxterms together to realize any Boolean function as well. We will call such a realization a **maxterm realization** of the function. Although different in appearance, the minterm and maxterm realizations of any given Boolean function are equivalent, as a check of their truth tables will show. The method we use to derive maxterms is a generalization of the style of thinking we used in the last chapter when we derived a realization of the implication function in terms of AND, OR, and NOT.

Consider the three-input Boolean function specified by the truth table shown in Table 5.11.

TABLE 5.11

x	y	z	
0	0	0	1
0	0	1	1
0	1	0	1
0	1	1	0
1	0	0	1
1	0	1	1
1	1	0	1
1	1	1	1

The minterm realization of this function (again, with AND symbols omitted) is $M_0 \vee M_1 \vee M_2 \vee M_4 \vee M_5 \vee M_6 \vee M_7$ or $\bar{x}\bar{y}\bar{z} \vee \bar{x}\bar{y}z \vee \bar{x}y\bar{z} \vee x\bar{y}\bar{z} \vee x\bar{y}z \vee xy\bar{z} \vee xyz$.

We can, however, come up with a better, more compact realization of the function. According to the truth table, the function is 0 only when $x = 0$ and $y = 1$ and $z = 1$. The function, then, is 1 whenever $x = 1$ or whenever $y = 0$ or whenever $z = 0$. In fact, it appears that the function might be realized by the expression $x \vee \bar{y} \vee \bar{z}$, because such a function would always be 1 except in the one case in which $(x, y, z) = (0, 1, 1)$ (Table 5.12).

TABLE 5.12

x	y	z	x	$\bar{y}$	$\bar{z}$	$x \vee \bar{y} \vee \bar{z}$
0	0	0	0	1	1	1
0	0	1	0	1	0	1
0	1	0	0	0	1	1
0	1	1	0	0	0	0
1	0	0	1	1	1	1
1	0	1	1	1	0	1
1	1	0	1	0	1	1
1	1	1	1	0	0	1

Indeed, $x \vee \bar{y} \vee \bar{z}$ is our function. The only row in the table in which x, $\bar{y}$, and $\bar{z}$ are all 0 is row three, in which $(x, y, z) = (0, 1, 1)$. An expression like $x \vee \bar{y} \vee \bar{z}$ is called a **maxterm,** short for *maximum term* because it is almost always 1, and only 0 for one particular combination of input bits. A maxterm consists of all the input variables of a given function, complemented or not, ORed together.

The general method for writing out the maxterm corresponding to a function that only has one 0 in the output column of its truth table is as follows:

1. Write all the variables in the function with ORs between them.
2. Draw a NOT bar over each variable *that is 1* in the single combination of inputs that makes the function 0.

Applying this method to our example, we first write down the input variables with ORs between them: $x \vee y \vee z$. Now we draw NOT bars over the inputs that were 1 in the one combination of inputs that made the function 0. In this case, that combination is $(x, y, z) = (0, 1, 1)$ in which both y and z are 1, so we draw NOT bars over y and z: $x \vee \bar{y} \vee \bar{z}$. The circuit diagram of this function is shown in Figure 5.9.

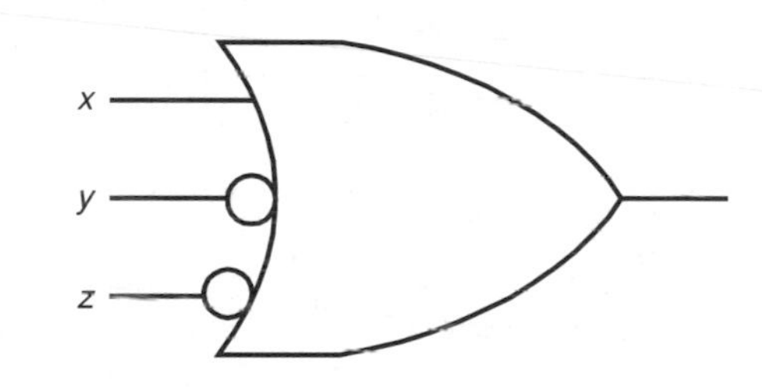

Figure 5.9 $x \vee \bar{y} \vee \bar{z}$.

We will revisit the troubling two-input OR function, which yielded the unwieldy minterm expression $(\bar{x} \wedge y) \vee (x \wedge \bar{y}) \vee (x \wedge y)$. We are now in a position to find the maxterm realization of the same two-input OR as shown in Table 5.13.

TABLE 5.13

x	y	$x \vee y$
0	0	0
0	1	1
1	0	1
1	1	1

There is only one 0 in the entire output column in row 0 of the truth table. First, we must write down the function's input variables with ORs between them: $x \vee y$. Now we draw NOT bars over any variables that are 1 in the combination of inputs that makes the function 0. In this case, the function is 0 only when $(x, y) = (0, 0)$, in which neither of the inputs is 1. Thus, we draw no NOT bars, and we are done. The maxterm realization of this function is the familiar $x \vee y$. This is not to say that the minterm realization of the OR function, $(\bar{x} \wedge y) \vee (x \wedge \bar{y}) \vee (x \wedge y)$, is incorrect; it is a perfectly valid realization of the same function. It is just that in this case the maxterm realization is simpler and thus easier to deal with algebraically.

5.6 REALIZING ANY BOOLEAN FUNCTION WITH MAXTERMS

Just as we found that we could OR minterms together to write the minterm realization for any Boolean function for which we have a truth table, we can AND maxterms together to write a maxterm realization for any Boolean function.

For example, consider the XOR function in Table 5.14.

TABLE 5.14

x	y	$x \oplus y$
0	0	0
0	1	1
1	0	1
1	1	0

Two of its four possible combinations of inputs cause it to be 0: the combinations $(x, y) = (0, 0)$ and $(x, y) = (1, 1)$. To write its maxterm realization, we write the maxterms for each of the combinations of inputs that produce 0's as output and write ANDs between them (circuit diagram shown in Fig. 5.10).

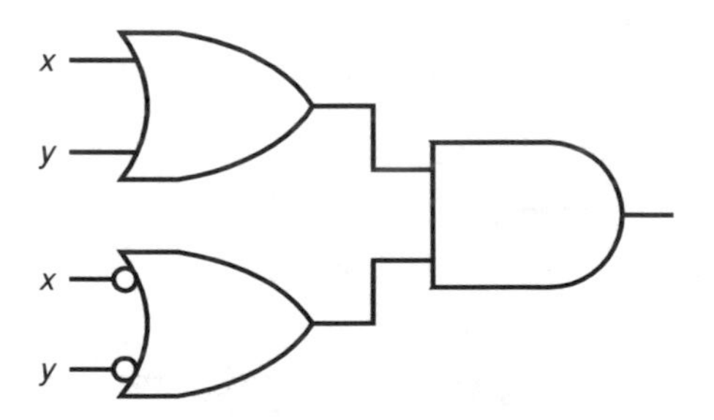

Figure 5.10 Maxterm realization of $x \oplus y$.

$$\text{maxterm for } (x, y) = (0, 0)\text{: } x \vee y,$$

$$\text{maxterm for } (x, y) = (1, 1)\text{: } (\bar{x} \vee \bar{y}),$$

$$\text{maxterm realization for XOR: } (x \vee y) \wedge (\bar{x} \vee \bar{y}).$$

The minterm realization of XOR, $(\bar{x} \wedge y) \vee (x \wedge \bar{y})$, essentially says that x XOR $y = 1$ when $x = 0$ AND $y = 1$ OR when $x = 1$ AND $y = 0$. The maxterm realization of x XOR y says that x XOR $y = 1$ when $x = 1$ OR $y = 1$ AND $x = 0$ OR $y = 0$; in other words, when at least one of the inputs is 0 and at least one of the inputs is 1. The two interpretations are just equivalent but different ways of looking at the same function. (Fig. 5.11).

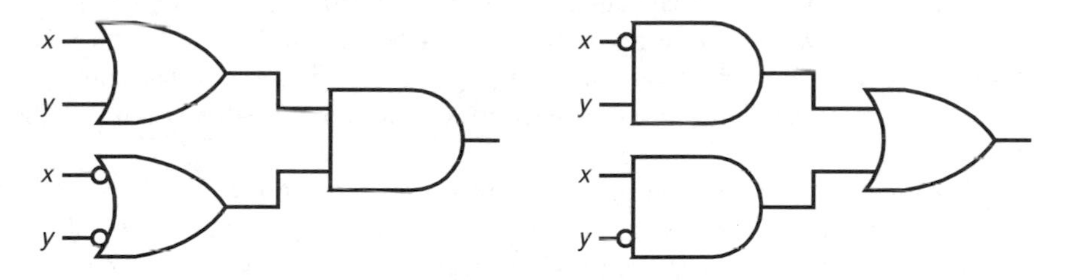

Figure 5.11 Two equivalent realizations of the XOR function: the maxterm realization (left) and minterm realization (right).

In general, if you want to write out an algebraic expression for a given truth table, and you must decide between realizing the function with minterms and realizing it with maxterms, it is easiest to use whichever method yields the shorter expression. To determine which realization this is, count the 1's and 0's in the output column of the truth table. If there are more 1's than 0's, use the maxterm realization, which has one maxterm for each 0 in the output column. If there are more 0's than 1's, realize the function with a minterm expression, which has one minterm for each 1 in the output column. If there is an equal number of 0's and 1's, use the method you feel more comfortable with.

TABLE 5.15 Procedure for Deriving Minterm and Maxterm Expressions

To Derive a Minterm Realization of a Function:	To Derive a Maxterm Realization of a Function:
• For a given row of the function's truth table that contains a **1** in the output column, write all input variables ANDed together.	• For a given row of the function's truth table that contains a **0** in the output column, write all input variables ORed together.
• Put NOT bars over those input variables that are **0** in that row.	• Put NOT bars over those input variables that are **1** in that row.
• Repeat this procedure for all rows in the truth table in which a **1** appears in the output column.	• Repeat this procedure for all rows in the truth table in which a **0** appears in the output column.
• **OR** the resulting **minterms** together.	• **AND** the resulting **maxterms** together.

5.7 PRODUCT-OF-SUMS EXPRESSIONS

As minterm expressions are examples of sum-of-product expressions, maxterm expressions are examples of **product-of-sums** (POS) expressions. POS expressions are expressions consisting of one or more sums ANDed together (a sum, in this case, being one or more variables, complemented or not complemented, ORed together). In a maxterm expression, however, each sum contains every variable present in the entire expression, while this is not necessarily true of all POS expressions. As with SOP expressions, if there is only one variable in a sum or only one sum in a product, the corresponding OR or AND is omitted. In the case of the maxterm realization of the OR function itself, for example, $x \vee y$, there is only one maxterm, so there is no need for an AND. As with SOP expressions, only individual variables are negated; an expression that contains the NOT function applied to groupings of more than one variable is not a POS expression.

The smallest unit of either a POS expression or an SOP expression is the variable. Both types of expressions at the lowest level are just made of variables linked by various operations. At a higher level, however, POS expressions are made up of sums (one or more variables, complemented or not, ORed together), while SOP expressions are made up of products (one or more variables, complemented or not, ANDed together). The name given to this intermediate unit of POS and SOP expressions is **term.** When we are dealing with POS expressions, a term is understood to mean a sum, and when we deal with SOP expressions, a term is understood to mean a product.

The circuit diagrams of all POS expressions in general and maxterm expressions in particular have a similar form (Fig. 5.12). They consist of one or more

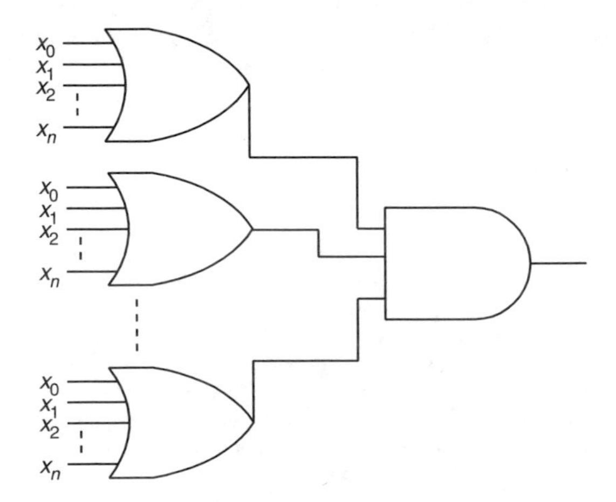

Figure 5.12 General circuit diagram of a function in POS form.

OR gates with variables as inputs and with outputs that all serve as inputs to one AND gate. The output of the function is the output of the AND gate.

Any of the inputs x_0–x_n to any of the OR gates in the diagram above may be negated as long as the negated inputs to any particular OR gate do not exactly match the negated inputs to any other OR gate (i.e., there are no redundant sums in the product of sums expression).

5.8 THE THREE-INPUT MAJORITY VOTER

Suppose you had to design a three-input majority voter; that is, a three-input Boolean function that counts its input bits. If there are more 1's than 0's, the function is to produce a 1 as output, and if there are more 0's than 1's, it is to produce a 0. Because there are three inputs, there can never be a tie. How would you realize this function?

First, we need a truth table. We will begin with the standard three-input truth table, with inputs named *x, y,* and *z* (Table 5.16).

TABLE 5.16

x	*y*	*z*	
0	0	0	
0	0	1	
0	1	0	
0	1	1	
1	0	0	
1	0	1	
1	1	0	
1	1	1	

Now we only need to fill in the output column of each row with the bit, 0 or 1, that tells the majority of input bits in that row (Table 5.17).

Counting the output bits reveals that there are four 0's and four 1's. Because it does not matter which method we choose to realize the function, we will use a maxterm realization. The rows that have 0's in the output column are $(x, y, z) = (0, 0, 0)$, $(x, y, z) = (0, 0, 1)$, $(x, y, z) = (0, 1, 0)$, and $(x, y, z) = (1, 0, 0)$. These are the rows whose input bits are either all 0 or have only one 1 among them. To find the maxterm realization of the three-input majority function, we find the maxterm corresponding to each of these four combinations of inputs and AND the resulting four maxterms together (Table 5.18).

TABLE 5.17

x	y	z	
0	0	0	0
0	0	1	0
0	1	0	0
0	1	1	1
1	0	0	0
1	0	1	1
1	1	0	1
1	1	1	1

TABLE 5.18 Individual Maxterms for Realization of Three-Input Majority Voter

Combination of Inputs	Corresponding Maxterm
$(x, y, z) = (0, 0, 0)$	$x \vee y \vee z$
$(x, y, z) = (0, 0, 1)$	$x \vee y \vee \bar{z}$
$(x, y, z) = (0, 1, 0)$	$x \vee \bar{y} \vee z$
$(x, y, z) = (1, 0, 0)$	$\bar{x} \vee y \vee z$

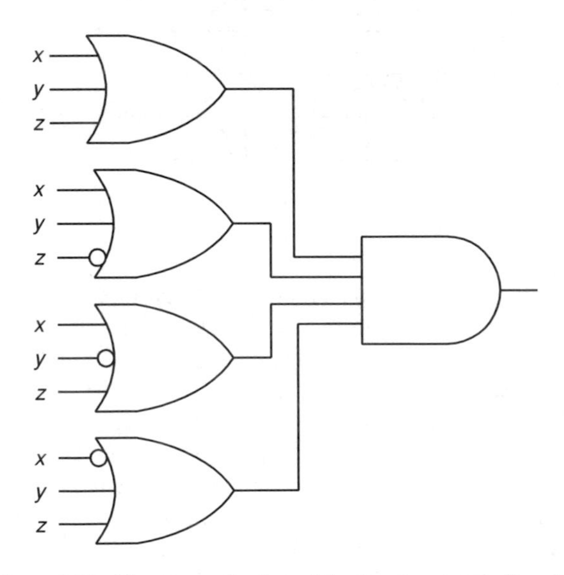

Figure 5.13 Maxterm realization of the three-input majority voter.

The final maxterm realization of this function then is $(x \vee y \vee z) \wedge (x \vee y \vee \bar{z}) \wedge (x \vee \bar{y} \vee z) \wedge (\bar{x} \vee y \vee z)$. Its circuit diagram is shown in Figure 5.13.

EXERCISE 5.2

A. Write the (a) minterm and (b) maxterm realizations of the following:

1. NAND
2. NOR
3. The implication function
4. AND

5–8. Write the (a) maxterm realization, (b) corresponding Venn diagram, and (c) circuit diagram of each of the functions specified by the truth tables given in exercises 1–4 in exercise set 5.1.

9. Write the minterm realization of the three-input majority voter.
10. Write the maxterm realizations of each of the seven functions needed to implement a full seven-segment display.
11. Write the Boolean expression(s) to implement a four-input majority voter. That is, a device with four input bits (a, b, c, and d) and two output bits called m (for majority) and t (for tie). If there are more 0's than 1's among the four input bits, m is to be 0 and t 0: if there are more 1's than 0's, m is to be 1 and t 0; if there is a tie (two of the inputs bits are 0 and two of them are 1) m is to be a don't care and t is to be 1. Should you use minterms or maxterms to realize m and t?

6

More Digital Circuits

In this chapter we will examine some more examples of Boolean algebra applied to digital circuit design. These circuits and the decoder introduced in chapter 5 are some of the fundamental building blocks of complex digital systems including computers. In addition to providing insight into the seemingly magical world of high technology that surrounds us, an exploration of the design of these circuits provides an opportunity to apply the techniques and principles we have learned about in previous chapters. The daunting term ''digital circuit design'' aside, the specific, hands-on focus of this chapter ought to be reassuringly down-to-earth after the abstractions we have covered up to now. Moreover, the approaches that we must take to design such circuits and the new ways in they force us to think about Boolean functions will broaden our understanding of Boolean algebra in a purely mathematical sense as well.

6.1 THE MULTIPLEXER: DATA VERSUS CONTROL

The first circuit that we will look at in this chapter is somewhat similar in conception to the decoder in chapter 5. It is called a **multiplexer,** or **MUX** for short. The multiplexer has one output line and two different kinds of inputs: data inputs and control inputs. For a number n (which depends on the desired size of the multiplexer) there are n control input bits and 2^n data input bits. The control bits (collectively called the **control signal**) choose which of the 2^n data input bits is to be patched through to the single output line in the same way the decoder input signal chose which output line would be 1.

For example, a MUX with a control signal consisting of two control bits (let us call them c_0 and c_1) would have $2^2 = 4$ data input lines (let us call them d_0, d_1, d_2, and d_3). There is exactly one data input bit for each of the four possible combinations of the two control bits, and the particular combination of control bits chooses which of the data bits will be sent through to the output line. Thus, the single output bit of such a multiplexer will be the same as data input d_0 if $(c_0, c_1) = (0, 0)$, d_1 if $(c_0, c_1) = (0, 1)$, d_2 if $(c_0, c_1) = (1, 0)$ and d_3 if $(c_0, c_1) = (1, 1)$. We do not know or care whether the data input bits (or the output bit) is 0 or 1 as long as the bit on the output line is the same as the bit on the data line we select with the control signal. If $(c_0, c_1) = (1, 0)$ and $d_2 = 1$, then the output bit is 1. If $(c_0, c_1) = (1, 0)$ and $d_2 = 0$, then the output bit is 0. Conceptually, the multiplexer functions like the device shown in Figure 6.1.

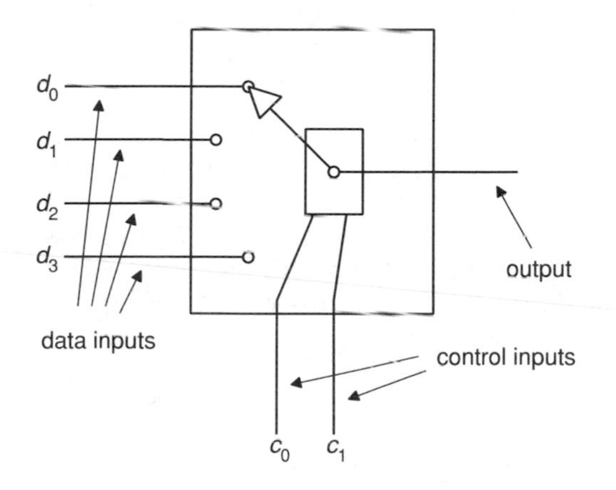

Figure 6.1 Conceptually, the multiplexer may be thought to contain an actual switch, under the control of the control inputs, that connects one of the data inputs to the output.

The multiplexer has many uses for electrical engineers. It may be thought of as a controllable data path. A particular combination of control bits may function as a command to let certain data through while withholding other data. Later we will see how this controllable aspect of the multiplexer allows it to be used to construct the number-crunching heart of a computer's **central processing unit** (CPU).

How can we build the multiplexer out of our favorite three functions? We could simply treat a four-data, two-control MUX as a Boolean function with six input bits and one output bit, write out its truth table, and derive a minterm or maxterm realization of it. However, at $2^6 = 64$ lines long, a six-input truth table would be unwieldy, to say the least. Fortunately, there is a way of thinking about this problem that allows us to break it down into manageable parts.

Although the data and control lines are all just bits, there is an important conceptual difference between them. The control bits never appear as output.

Rather, they steer the whole circuit inside, controlling its operation. The data bits, however, are simply herded like cattle through the circuit. They are the raw material that is processed or filtered by the circuit under the direction of the control signal. The somewhat abstract distinction between the two different kinds of data, data-as-data and data-as-control, is crucial to digital circuit design and is the key to the construction of the multiplexer.

6.1.1 AND as Controllable Pass-Through Gate

To understand this distinction, we will take a momentary detour away from the multiplexer and revisit the simple two-input AND gate (Fig. 6.2).

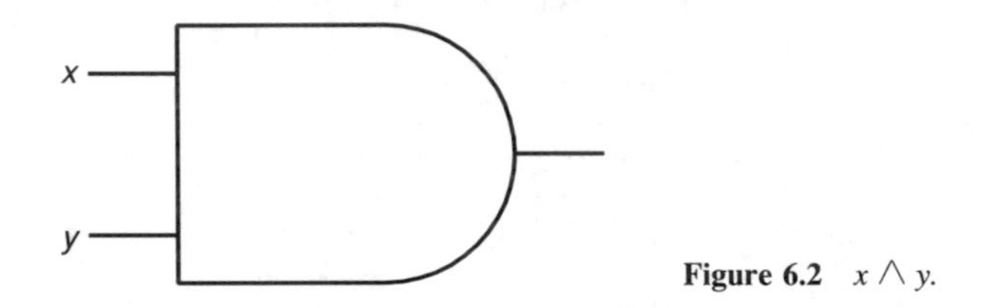

Figure 6.2 $x \wedge y$.

By arbitrarily deciding that one of the AND gate's inputs bits (x) is a control input and the other of its inputs (y) is a data input, we find that we have created a simple controllable data path in which y is the bit being controlled and x is the bit doing the controlling (Fig. 6.3).

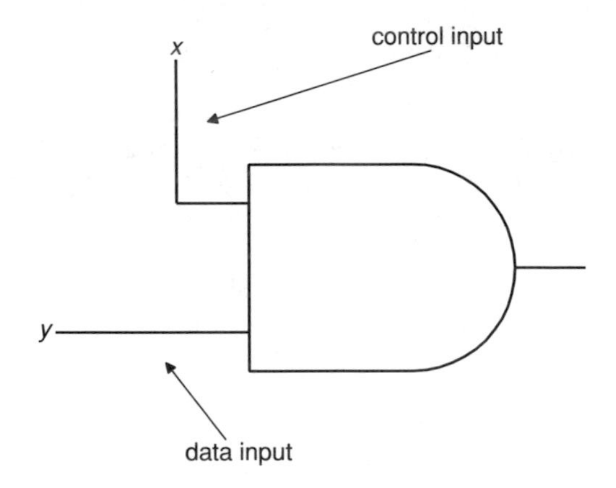

Figure 6.3 $x \wedge y$, with x arbitrarily designated the control input and y the data input.

How does x exert control over the output? Look at the truth table for AND, paying particular attention to the rows in which $x = 0$ (Table 6.1).Note that when $x = 0$, the output of the AND gate is 0 as well. However, consider what happens when $x = 1$ (Table 6.2).

When $x = 1$, the output bit becomes whatever y is. Thus the AND gate acts

TABLE 6.1

x	y	$x \wedge y$
0	0	0
0	1	0
1	0	0
1	1	1

TABLE 6.2

x	y	$x \wedge y$
0	0	0
0	1	0
1	0	0
1	1	1

as a controllable gate for data input y. If $x = 0$, the gate is shut off like a faucet and only 0 comes out no matter what y is. If $x = 1$, the gate simply passes y through unchanged. When an input is used to control another input in this way, we often speak, for example, of x **enabling** y (in the case in which $x = 1$) or **disabling** y (as in the case in which $x = 0$). Because AND is symmetric, we could just as easily have declared y to be the control bit and x to be the data bit. It is just a matter of perspective as a review of the truth table for AND will show. We will employ this ''controllable'' aspect of AND to realize our multiplexer.

6.1.2 Decoder-Based Realization of the Multiplexer

A possible approach to the design of the multiplexer would be to feed the multiplexer's control signals into a decoder. Each output line from the decoder could then be ANDed with a different data bit, and all the resulting bits ORed together. This solution is shown in Figure 6.4.

For each particular combination of control bits, only one of the decoder's output lines will be 1: the other three will be 0. The ''on'' decoder output line will then enable only the data line that it gets ANDed with; the other three data lines will be ANDed with 0's, and thus be disabled. Finally, the OR gate will receive as input three 0's and the enabled data bit. It will pass the data through, since ORing any bit with any number of 0's has no effect; the data bit is simply passed along unchanged as the output bit of the OR. Thus we have achieved our desired goal: we can select any one of the four data lines with our two control lines, and the output of the entire circuit is the bit being carried on the selected

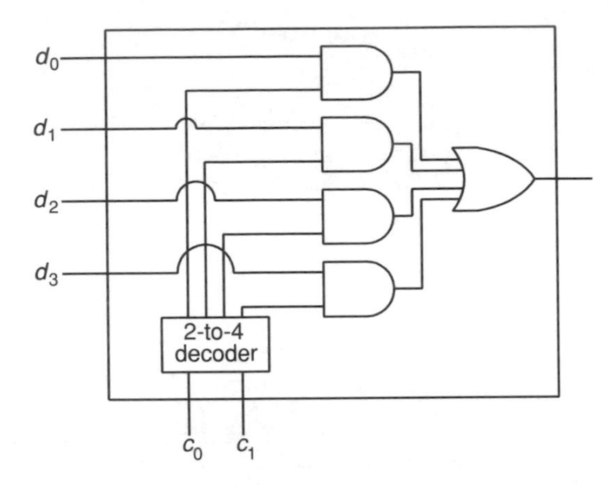

Figure 6.4 Decoder-based realization of 4-data, 2-control multiplexer.

data line. For example, if $(c_0, c_1) = (0, 1)$, then whatever data input d_1 is will appear on the output line as shown in Figure 6.5.

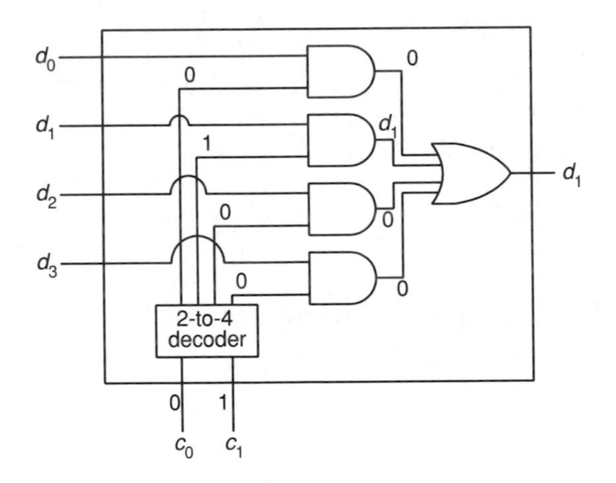

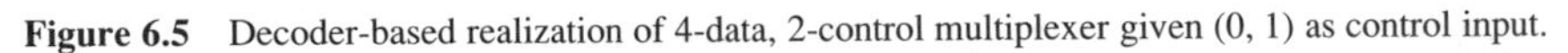

Figure 6.5 Decoder-based realization of 4-data, 2-control multiplexer given (0, 1) as control input.

However, each output line of a decoder is realized by the minterm of the particular combination of input bits that makes that output line 1. Thus, the above realization of the multiplexer could be drawn more specifically as shown in Figure 6.6.

6.1.3 Multiplexer with the Decoder Built In

In the decoder-based realization of the multiplexer described in Section 6.1.2, each input data bit goes through two levels of ANDs before it gets to the final OR. Thinking literally about what the multiplexer does, however, we can combine

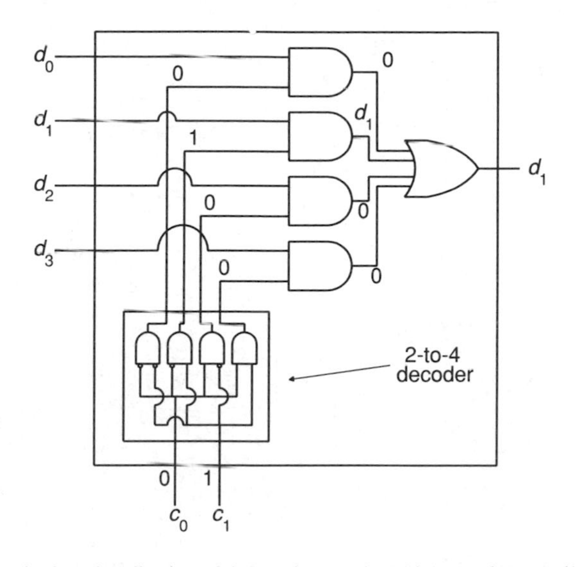

Figure 6.6 Decoder-based realization of 4-data, 2-control multiplexer given (0, 1) as control input with realization of decoder shown.

the logic in the decoder with the rest of the multiplexer circuit, thereby simplifying things a bit.

We want a circuit that yields d_0 as output when $\bar{c}_0$ AND $\bar{c}_1$, d_1 when $\bar{c}_0$ AND c_1, d_2 when c_0 AND $\bar{c}_1$, and d_3 when c_0 AND c_1. Just writing it out in this way points strongly to the final function. Indeed, the circuit we are looking for is realized in the following expression:

$$(d_0 \wedge \bar{c}_0 \wedge \bar{c}_1) \vee (d_1 \wedge \bar{c}_0 \wedge c_1) \vee (d_2 \wedge c_0 \wedge \bar{c}_1) \vee (d_3 \wedge c_0 \wedge c_1).$$

We simply ANDed each data input with the particular combination of control inputs that we wanted to select that data input and ORed all the results together. Effectively, this realization of the multiplexer has a decoder built in. The different combinations of c_0 and c_1 ensure that only one of the four ORed together expressions will ever be 1 at any given time. For example, if $(c_0, c_1) = (0, 1)$, of the four expressions above that are ORed together, all would be 0 except the second, $(d_1 \wedge \bar{c}_0 \wedge c_1)$. The combination of $(c_0, c_1) = (0, 1)$ would enable the d_1 in this expression, which would be passed alone to the OR along with three 0's. The output of the entire expression would then be the result of the three 0's ORed with d_1, as shown in Figure 6.7.

6.1.4 Realizing Any Boolean Function with a Multiplexer

In chapter 5 we learned how to use a decoder to realize any Boolean function in addition to the minterm and maxterm realization methods. Now we will examine a way of realizing any Boolean function using a multiplexer.

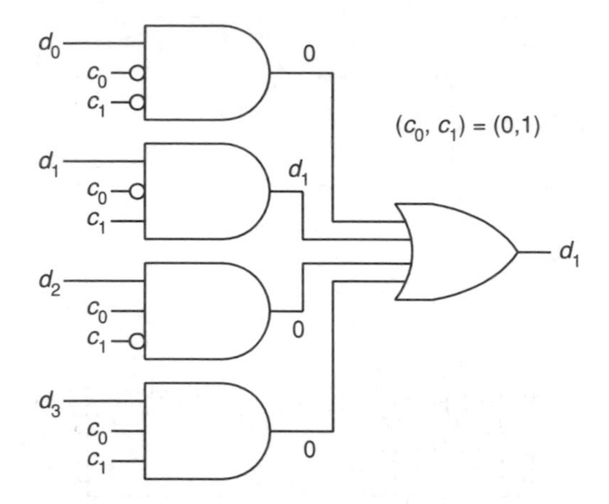

Figure 6.7 Simpler realization of 4-data, 2-control multiplexer given (0, 1) as control input.

Consider the three-bit Boolean function shown in Table 6.3. Instead of using minterms (or in this case, more likely maxterms) to realize this function, we can use a three-control, eight-data multiplexer. We simply feed the inputs of the desired function into the control inputs of the multiplexer so that each combination of input bits (x, y, z) chooses a different one of the multiplexer's eight data lines. Then we "hardwire" the data lines to be the proper output bits for each combination of inputs bits exactly as they appear in the output column of the truth table, resulting in the circuit shown in Figure 6.8.

The data lines shown in the multiplexer in Figure 6.8 do not have variable inputs. They are hardwired, meaning that they are permanently set to be a particular value. Specifically, each data input bit is set to be the desired output from the function for a different combination of inputs. For instance, according to the

TABLE 6.3

x	y	z	
0	0	0	1
0	0	1	0
0	1	0	0
0	1	1	1
1	0	0	0
1	0	1	1
1	1	0	1
1	1	1	1

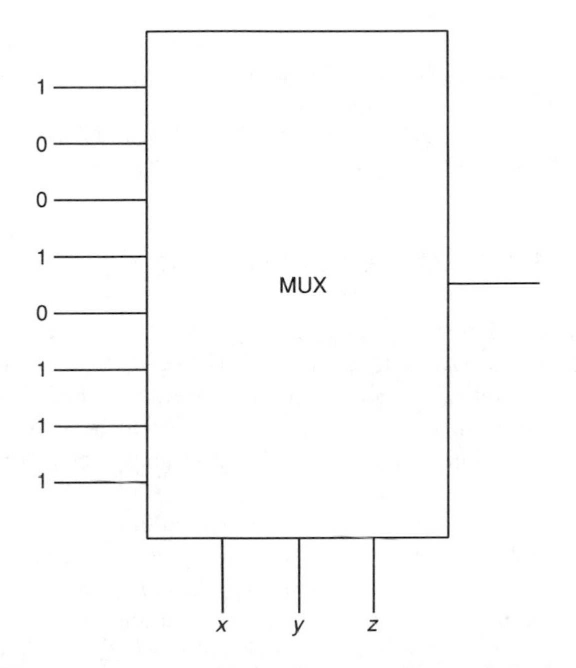

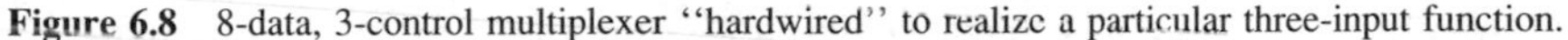

Figure 6.8 8-data, 3-control multiplexer "hardwired" to realize a particular three-input function.

truth table, when the function's inputs are $(x, y, z) = (1, 0, 1)$, the output of the function should be 1. In the circuit shown in Figure 6.8, when $(x, y, z) = (1, 0, 1)$, data line five is selected to be the output from the multiplexer. We have hardwired data line five always to be a 1, so when $(x, y, z) = (1, 0, 1)$, the output of the circuit is a 1. This method can be used for any Boolean function. Given any n-bit function, its truth table will be 2^n bits long, and we would need an n-control, 2^n-data multiplexer to realize it in this way.

This is a rather unintelligent, or "brute force," use of a multiplexer. By hardwiring the data inputs just the way we want them for the particular function we want to realize, we have just trained the multiplexer to memorize the truth table and look up the output bit for the row specified by a given value of the control signal. In this sense, however, we have implemented a sort of primitive computer memory that allows us access to the bit stored at a given **memory address** as specified by the multiplexer's control signal. It is, however, a read-only memory; we cannot change the bit at a given address, we can only read it.

EXERCISE 6.1

1. Draw a circuit diagram and write the corresponding algebraic expression for a three-control, eight-data multiplexer.

2. Derive full maxterm and minterm realizations of a one-control, two-data multiplexer.
3. Recalling our interpretation of the AND gate as a controllable pass-through gate, examine and describe in your own words how the two-input OR, NAND, NOR, XOR, and COIN behave when one input is considered a control input and the other a data input. What is the effect on the data input when the control input is turned on and off?

6.2 VECTORS AND PARALLEL OPERATIONS

Thus far we have dealt only with data that consisted of a single bit. The multiplexers we have looked at take individual data bits in and switch between them on the basis of the control signal. Sometimes, however, we want to switch between binary numbers that are many bits long. This is especially true in computer circuitry in which data is almost always considered in groups of 8, 16, 32, or more bits at a time.

For instance, suppose we wanted to build a multiplexer that selected one of four eight-bit numbers on the basis of a control signal. Because there are still only four things we are choosing among, we only need two bits of control to select one of them uniquely, even though each data input consists of more than one bit. Conceptually, this case is no different from the single-bit data case. However, at the logic gate level, it is slightly more complex.

Each of the four data input signals is to be eight bits wide, as is the output signal. That is, each of these **data paths** consists of eight bits going side by side into (or out of) the multiplexer. Bits that travel together and are meant to be interpreted as a single multi-bit number are bits that are taken **in parallel** (this parallelism of data transmission is not to be confused with that meant when we speak of parallel-processing computers).

We describe data paths as being a certain number of bits ''wide'' in the same sense that we speak of highways being four lanes wide. The width of the data path refers to its capacity in parallel bits. In diagrams of digital circuits, we indicate single-bit paths with the single arrow or the line that we have been using all along, and indicate multi-bit data paths with double-wide arrows. Sometimes a slash across the data path tells how many bits in parallel the data path accommodates.

Sometimes values consisting of many bits taken in parallel are known as **vectors.** Furthermore, single bits of data are usually represented with lowercase letters, while uppercase letters indicate multi-bit vectors. A high-level diagram of the multiplexer with eight-bit data paths is shown in Figure 6.9.

Such a diagram is called a **block diagram.** It depicts a circuit abstractly as a block or box, as well as its inputs and outputs, but does not bother with the details of the circuit's implementation in terms of AND, OR, and NOT gates. We will use block diagrams throughout this chapter to illustrate the different common circuits we will discuss.

How would we implement this vector-input multiplexer circuit using the

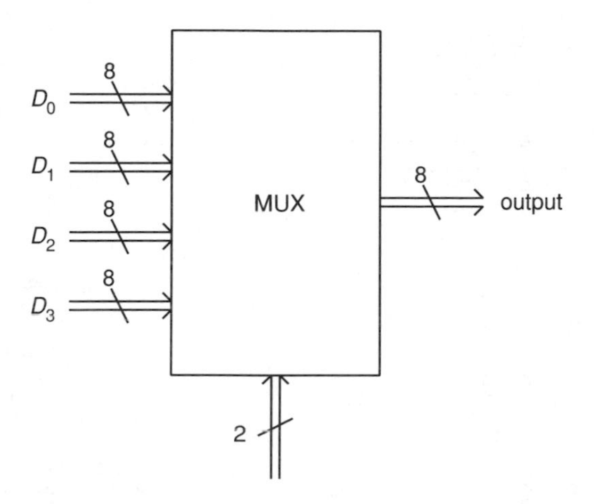

Figure 6.9 4-data, 2-control multiplexer in which data paths are eight-bit wide vectors.

circuits we already know about? The bits in a particular data vector only interact with the corresponding bits in other data vectors and not with each other. Therefore, to build a multiplexer with four eight-bit vectors as data inputs, we would build eight multiplexers that each take four single bits as data inputs and feed each of these eight multiplexers the same control signal. Each of the eight multiplexers (we will call them MUX0 through MUX7) handles one bit for the multi-bit multiplexer. MUX0 will take the first bit of each of the four eight-bit input data vectors as its four data inputs, MUX1 will take the second bit of each eight-bit vector as its inputs, and so on. Each of the eight multiplexers produces a single output bit. Taken together as an eight-bit number, these eight bits are the desired result of our eight-bit multiplexer.

The fourth such single-bit-input multiplexer in such an arrangement, MUX3, is shown in Figure 6.10. Note that it takes as its four single-bit inputs the fourth bit of each of the four eight-bit data vectors, D_0–D_3. The bit it produces as output is taken to be the fourth bit in the eight-bit output vector of the vector multiplexer.

Constructing multi-bit circuits in this way from single-bit circuits (or multi-bit circuits from smaller-capacity multi-bit circuits) is known as **bit slicing.** All vector operations are split apart into single-bit operations that are carried out in parallel, and the single-bit results are put together in parallel to form an output vector. Each single-bit circuit constitutes a slice of the final multi-bit circuit.

Any Boolean operation we have performed on single bits also can be performed on vectors in parallel. For example, to form an eight-bit vector that is the result of a parallel AND performed on two eight-bit vectors, simply AND the first bit of the two vectors together to form the first bit of the result, then AND the second bit of the two vectors together to form the second bit of the result, and so on.

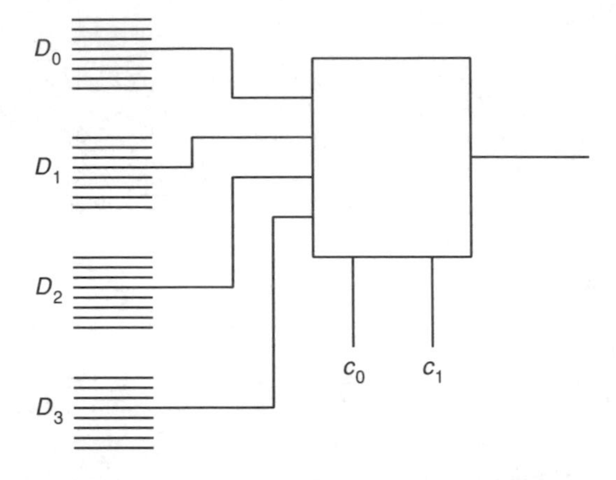

Figure 6.10 Block diagram of the circuit that produces the fourth output bit of the output vector from a 4-data/2-control multiplexer that operates on eight-bit data vectors.

It is important to realize that by allowing ANDs and other Boolean functions to perform parallel operations, we are not redefining them as much as we are using a shorthand so we do not have to write out the more confusing reality of the situation in which there are several single-bit path AND gates operating in parallel as shown in Figure 6.11.

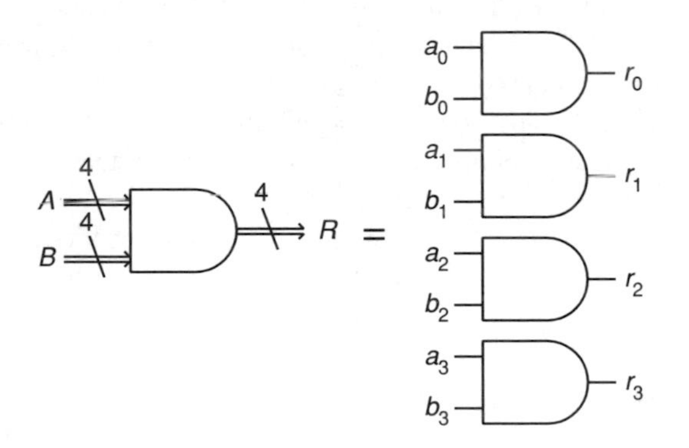

Figure 6.11 Four-bit vector AND: vector representation and single-bit reality. *R* is the output vector (*R* for ''result'').

A simple Boolean operation performed in parallel on each bit in multi-bit vectors is known as a **bitwise** operation. Figure 6.11 shows a circuit that performs a four-bit bitwise AND. Any time we use diagrams in which double arrows (multi-bit vectors) are shown going into and coming out of a Boolean gate, it is understood that the output vector is the result of the Boolean operation being performed on the input vector(s) in a bitwise or parallel fashion.

6.3 THE ADDER

How could we use the techniques we have learned to create a function that would add two eight-bit numbers in base 2? Such a function would have to have enough input bits to accommodate both eight-bit input numbers, or 16 bits. Just to write the truth table for such a function would involve $2^{16} = 65{,}536$ lines.

How many output bits would the function have? To answer this, we need to know the largest number that can result from the addition of two eight-bit numbers. The largest eight-bit number is $11111111_2 = 255$, so the largest sum of two eight-bit numbers we could possibly have is $255 + 255 = 510$, or 111111110_2, a nine-bit number. So the adder must have nine output bits (each a separate function itself), each having a truth table with 65,536 lines—hardly an inviting prospect. However, if we look closely at the mechanics of adding base-2 numbers together, we will find an easier way to build our adder.

To create a Boolean function that adds two base-2 numbers together, we must develop an algorithm for adding in base 2. That is, we need to determine a specific sequence of steps for adding two base-2 numbers. It is useful at this point to take a hard look at the method we use when we add two base-10 numbers together.

6.3.1 Adding in Base 10

We begin by writing one of the numbers above the other, with a line underneath, below which we will eventually write the sum:

$$\begin{array}{r} 539712 \\ \underline{2573} \end{array}.$$

We then proceed from right to left (least significant digit to most significant digit) adding the numerals that appear in successively more significant vertical columns. As long as the sum of the column of digits for any particular digit position is only one digit large itself, we simply write the sum digit below the line in the same vertical column as the digits summed:

$$\begin{array}{r} 539712 \\ \underline{2573} \\ 85 \end{array}.$$

However, we have a problem when we get to a column in which the digits being added yield a sum that is greater than can fit in one digit, as is the case in the next column to be added in our example (the third column). We cannot put the actual sum of 7 and 5, i.e., 12, in the same column beneath the 7 and the 5. Instead, we split the sum (12) up into a 10s digit (1) and a ones digit (2). We write the ones digit in the column beneath the 7 and the 5 in the same column and **carry** the 10s digit into the next column, where it is simply added in with the numbers already in that column. Thus, when that column is added up, there are actually three numbers that are summed: the original two digits in the column and the digit carried over from the column to its right:

$$\begin{array}{r} 1 \\ 539712 \\ \underline{2573} \\ 285. \end{array}$$

This process is repeated until we run out of digits in our numbers being added. To make the algorithm simpler, we could say that there is always a carry-over into the next column, but that the carry is sometimes zero, as when we add 2 and 3 (5 in the ones digit written below and 0 in the tens digit, carried into the next column). According to this way of thinking about addition, every column is actually the sum of three numbers. We could even say that the first column, the one on the far right, automatically has a zero added in to it as the carry-over from the (nonexistent) column to its immediate right.

What happens when one number runs out of digits before the other one does? We continue adding as before, but we **pad** the shorter number with zeros to the left of its most significant digit out to the length of the longer number. Here then is the same addition, but with the ''invisible'' padding and carry zeros shown:

$$\begin{array}{r} 011000 \\ 539712 \\ \underline{002573} \\ 542285. \end{array}$$

6.3.2 Adding in Base 2

We use essentially the same method to add base two numbers together. For example, let us add the base-2 numbers 111001 (57_{10}) and 10011010 (154_{10}). By adding 57 and 154 in base 10, we already know that the answer should be 211 (11010011_2). As before, we begin by writing the two numbers to be added, one above the other with a line beneath them. We may as well also take this opportunity to pad the shorter number out to eight bits with zeros:

$$\begin{array}{r} 10011010 \\ \underline{00111001}. \end{array}$$

Also as before, we add the numbers vertically, beginning with the rightmost column. (Keep in mind that for the purposes of this addition, we are thinking of the ones and zeros involved as integers, which are subject to the rules and functions of ordinary arithmetic, and not as the logical bits we have been using in our Boolean expressions. We are, however, adding in base 2 instead of base 10, so the arithmetic may not be completely ordinary after all.) The addition of two base-2 digits is relatively simple: $0 + 0 = 0$, $0 + 1 = 1$, $1 + 0 = 1$, and $1 + 1 = 10$. It all works just as it does in base 10, except the last case, $1 + 1 = 10$, in which the sum is in base 2 ($10_2 = 2_{10}$):

$$\begin{array}{r} 10011010 \\ \underline{00111001} \\ 011. \end{array}$$

However, we now get to a column in which the sum of the digits will not fit in the single digit beneath the column: $1 + 1 = 10$. We handle this in a way that is analogous to the way we handled the corresponding base-10 case. We break the sum into a ones digit (0) and a twos digit (1). We write the ones digit in the proper column beneath the sum line and carry the twos digit into the next column:

$$\begin{array}{r} 1 \\ 10011010 \\ \underline{00111001} \\ 0011. \end{array}$$

We now have a case in which we are adding three ones together. In this case $1 + 1 + 1 = 11$, so we write the ones digit (1) beneath the line and carry the twos digit (1) over to the next column. Continuing in this way, we eventually have

$$\begin{array}{r} 111 \\ 10011010 \\ \underline{00111001} \\ 11010011. \end{array}$$

Just as we predicted, the sum is 11010011, or 211_{10}. As with addition in base 10, we may as well assume that there is always a carry digit from the previous column when we add the digits in a column together even if this carry digit is zero. This helps us standardize the operation we perform in each digit column. In each column, then, we always add three digits, and we always produce two digits (the sum digit, which goes below the sum line, and the carry digit to be added in the next column even if that carry is zero).

6.3.3 The Binary Adder Function

Because we find ourselves performing the same operation over and over again on successive columns of three digits each, whenever we add long numbers together, for our binary adder function we will first build a function that will just deal with one column. We then will repeat the same function as many times as we want, cascading these identical functions together, depending on how large the numbers are that we want to add together. When approached in this way, the problem becomes much simpler than it first appeared.

The adder function that we build for a single column is to have three inputs: the two bits to be added (one from each of the multi-bit numbers we are adding together) and a carry bit from the previous column. There are to be two output bits: the sum bit and the carry bit to be fed into the next column. Thus each single-column adder is actually two functions, each using the same input bits: the sum function and the carry function. The sum bit is that single-column adder's contribution to the final sum vector. The carry bit is the link between adjacent one-bit adder functions. Each column's carry output bit is connected to the next column's carry input bit. The first column's carry input bit is always fed a 0, and the last column's output carry bit is simply used as the most significant sum bit. The block diagram of an n-bit adder is shown in Figure 6.12.

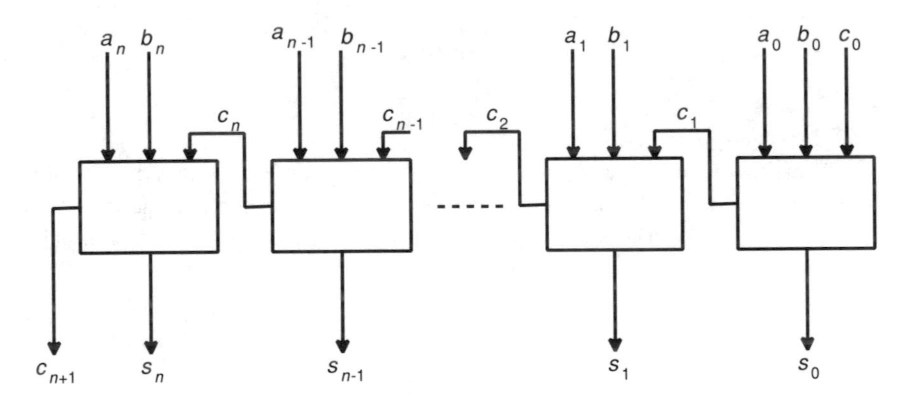

Figure 6.12 Block diagram of a circuit that adds two n-bit numbers, A and B. The resulting sum is $n + 1$ bits long.

The adder circuit is an example of a cascading circuit because it is made of a series of identical simple modules or cells with bits cascading from one to the next. It cannot be built in a parallel fashion (not easily, anyway) as the the multi-bit multiplexer was, with each bit going through a single bit adder and ignoring the other additions going on at the same time. This is because each column's addition depends on an adjacent column's addition for its carry input bit; each cell interacts with its neighbors. This is what distinguishes a cascading circuit from a parallel circuit.

In the real world of electrical engineering and circuit design, electricity does not propagate instantaneously through circuitry. There is a very real penalty incurred by cascading circuits in terms of the time it takes the binary signals to ripple their way through the circuit before a final, accurate output is produced. Naturally, this delay increases as the number of circuits being cascaded increases. While there are clever techniques that engineers use to defeat or mitigate this delay, once noted, they are beyond the scope of this book.

From here, building the adder is relatively straightforward. Once we have functions capable of generating the sum and carry bits for a single column, we can replicate them as many times as necessary (depending on the size of the numbers we want to be able to add), cascading all of the copies together. To realize the single-bit sum and carry functions, we need a truth table for each.

Because the sum and carry functions each take three bits as input, their truth tables will be eight lines long. Since they both operate on the same inputs, we may as well write their output columns on the same truth table, side by side. We will call the two input bits from the numbers to be added a and b (it does not matter which is the bit from the top number and which is the bit from the bottom—the sum and carry are the same either way). We will call the input carry bit c_i for *carry in*. The two output bits, the sum bit and the carry output bit, will be called s and c_o, respectively.

If we put the c_o output column to the left of the s output column and consider the two of them together as a two-bit binary number, then our truth table is simply an addition table in base 2. The value of the vector (c_o, s) in each row is just the two-bit sum of the three individual bits in the input columns in that row (Table 6.4).

TABLE 6.4

a	b	c_i	c_o	s
0	0	0	0	0
0	0	1	0	1
0	1	0	0	1
0	1	1	1	0
1	0	0	0	1
1	0	1	1	0
1	1	0	1	0
1	1	1	1	1

Using the truth table shown in Table 6.4 and the techniques for realizing any Boolean function, it should now be simple to derive the actual minterm or maxterm realizations of the adder and carry functions. This is left as an exercise for the reader.

6.4 THE COMPARATOR

Another useful cascaded circuit is the *comparator.* It compares two binary numbers and determines which one is larger (or whether or not they are equal, if we chose to build that capability into the comparator). As with the adder, we build a comparator by cascading any number of identical circuits together with the number of circuits we cascade in this way being the sole limit on the length of the binary numbers we may compare. As with the adder, we assume that both input numbers (let us call them A and B) are padded out to the maximum length with padding zeros.

Before we can dive in and start writing truth tables, we need to decide exactly what the comparator should do. Whereas the adder took two n-bit binary numbers as input and produced an $(n + 1)$-bit number as a result (with each cell dropping a single bit into the final result, except the final cell, which produced two of the sum bits), the comparator does not produce a vector. How many bits should the entire circuit produce as output? How much information do we want the circuit to deliver?

Let us say that we want our comparator to tell us which number is larger if the numbers are not equal and also whether or not they are equal. This comes down to three possible states of affairs between which our comparator must distinguish: $A = B$, $A < B$, and $A > B$. We know that we need at least two bits to represent three different conditions, so we will have two output bits for the entire comparator, and we will call them e (for "equal") and a_g (for "A is greater"). If e is 1, then $A = B$, and a_g must be 0 (the combination $(e, a_g) = (1, 1)$ will not be produced by our comparator). The combination $(e, a_g) = (0, 1)$ is to mean that $A > B$, and the combination $(e, a_g) = (0, 0)$ is to mean that $A < B$.

Let us think for a moment about how we compare two multi-digit numbers A and B in any base. If we start at the most significant digit position, we look at the numeral in that position in A and the numeral in the same position in B. If they are not equal, we have our answer right there; whichever of A and B has the bigger numeral in the most significant digit position is the bigger number, period, no matter what the rest of the numerals in the other digit positions are in either number. As soon as you look at the 4 in 403,961 and compare it to the 3 in 390,265, you know that 403,961 is greater than 390,265.

If, however, the numerals in the most significant digit positions in A and B are the same, we must look at the next digit over. If the numerals in this digit position are then unequal, once again we have our final answer, and if they are equal, we move once more to the right to the next digit position after that. If we make it all the way to the final, least significant digit position and A and B have been equal in all digit positions up to and including that point, then $A = B$.

To create a comparator as a cascaded circuit, we need to create a single-column comparator, then chain any number of these together. The single-column comparator should take four bits in, a and b (the two bits from A and B, respectively that correspond to the bit position that this single-column comparator is concerned with); e_i (equal in or "A and B are equal so far"); and a_{gi} (A is greater in or "A has already been determined to be greater than B"). The bits e_i and a_{gi} come from the cell to the immediate left, i.e., from the cell whose bit position is more significant than that of the current cell (from upstream). If the single-column comparator gets $(e_i, a_{gi}) = (0, 1)$, this means that A has been determined somewhere upstream to be greater than B, whereas if $(e_i, a_{gi}) = (0, 0)$, this means that B has been determined to be greater than A. $(e_i, a_{gi}) = (1, 0)$ means that A and B are equal in all bit positions looked at so far (i.e., to the left of, or upstream from, this bit position).

Our single-column comparator is to produce e_{go} (equal out or still equal so far) and a_{go} (A is greater out) as output to be fed dowstream to the cell to the immediate right. Because the single-column comparator produces two bits as output (like the single column adder that produced both a sum and a carry bit), it is really two separate functions, each acting on the four input bits.

Because we may think of the numbers being compared as being padded out to infinity with zeros, they are to be considered equal up to the point where the comparator starts looking at them, i.e., at the most significant bit. Thus we feed $(e_i, a_{gi}) = (1, 0)$ into the leftmost cell of the comparator to start things off.

Unlike the single-bit adder cells that made up the adder circuit, the cells in the comparator do not contribute directly to the output of the entire circuit. Rather, they will only feed each other, passing signals down the line, and the final cell's outputs will be taken as the output of the whole comparator. A block diagram of the comparator is shown in Figure 6.13.

How do the two functions in a single-column comparator operate on their four input bits (a, b, e_i, a_{gi}) to produce e_o and a_{go}? We could write out two 16-row truth tables and work through all the combinations, but common sense and the brief analysis of comparison above suggest a simpler way.

Let us first make the function to generate e_o for a given bit position, since it is quite easy. Basically, if the numbers being compared are equal so far (i.e., e_i is 1 coming into this cell) and a is equal to b, then e_o is 1. In all other cases (there is at least one inequality in a bit position somewhere upstream, or a and b are not equal in this bit position), e_o is to be 0. Thus the function to generate e_o is simply:

$$e_o = e_i \wedge (a \odot b).$$

Or in terms of AND, OR, and NOT:

$$e_o = e_i \wedge [(\bar{a} \wedge \bar{b}) \vee (a \wedge b)].$$

The circuit diagram for this function is shown in Figure 6.14.

For the function to generate a_{go}, we use the following reasoning. For any given bit position, if e_i is 0, then it has already been determined by a cell in a

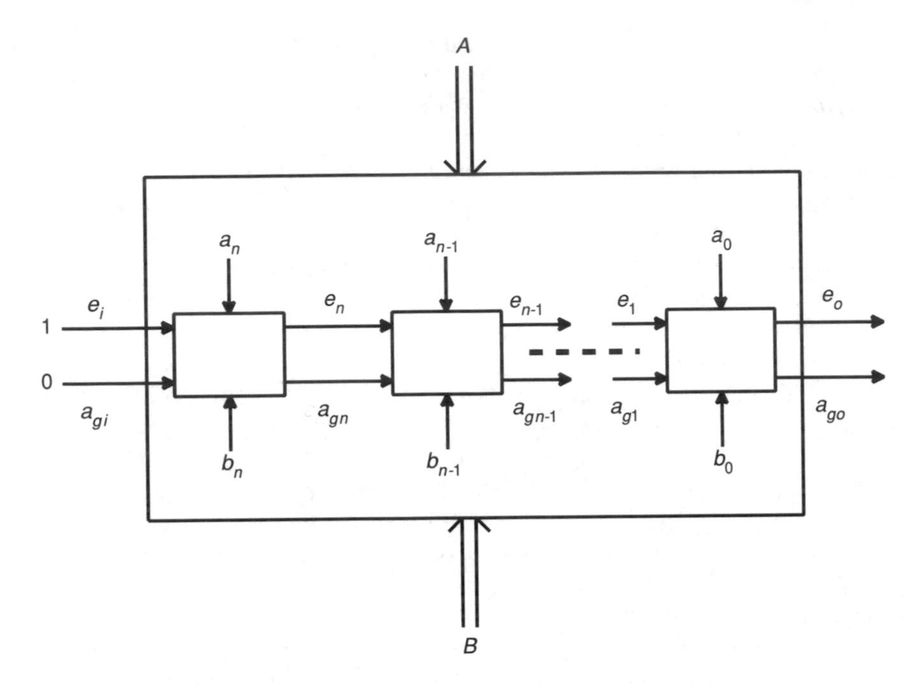

Figure 6.13 Block diagram of an n-bit comparator.

more significant bit position which of A and B is greater, and no value of this column's a and b could have any effect on that determination. In this case, we simply pass the information given to us in a_{gi} on through to the next cell as our a_{go} output. If, however, e_i is 1, then it may be up to this cell to determine which is greater. If e_i is 1 and a equals b, then A equals B so far to our left, and A still equals B even after we have examined our small part of A and B, namely a and b. In this case we pass out $(e_o, a_{go}) = (1, 0)$, and so a_{go} is to be 0. If, however, e_i is 1 and a does not equal b, then this bit position is the one that determines which of A and B is greater. This situation breaks down into two distinct cases. If $(a, b) = (0, 1)$, then a_{go} is to be 0 because then we will know that B is greater

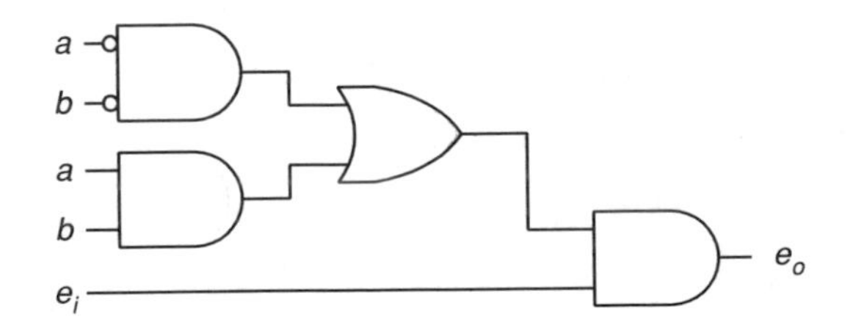

Figure 6.14 Function to realize e_o of one cell of an n-bit comparator.

than A; and if $(a, b) = (1, 0)$, then a_{go} is to be 1 because we will then know that A is greater than B.

Given all of that, when exactly should a_{go} be 0 and when should it be 1? If e_i is 0, then a_{go} should be whatever a_{gi} is; so it should be 1 if e_i is 0 and a_{gi} is 1. However, we know that a_{gi} will never be 1 unless e_i is 0, so we can simply say that a_{go} should be 1 whenever a_{gi} is 1. Output a_{go} should also be 1 if e_i is 1 and $(a, b) = (1, 0)$. It should be 0 in all other cases. If a_{go} is to be 1 if and only if $a_{gi} = 1$ or ($e_i = 1$ and $a = 1$ and $b = 0$), our final realization of the function to generate a_{go} is:

$$a_{go} = a_{gi} \vee [e_i \wedge (a \wedge \bar{b})].$$

We may use a three-input AND to get rid of one layer of parentheses without changing the value of the expression:

$$a_{go} = a_{gi} \vee (e_i \wedge a \wedge \bar{b}).$$

The circuit diagram for this function is shown in Figure 6.15.

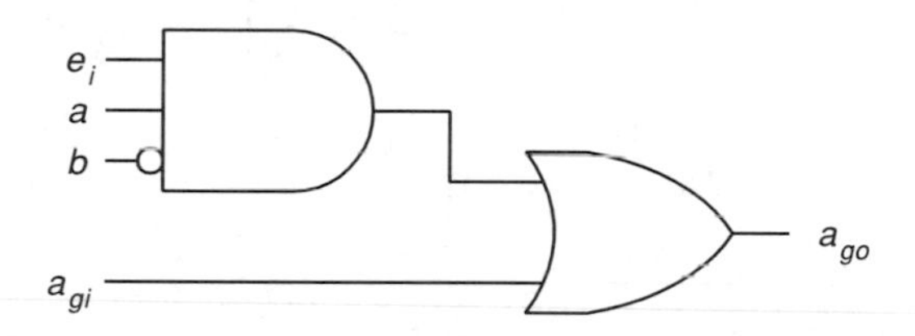

Figure 6.15 Function to realize a_{go} of one cell of an n-bit comparator.

EXERCISE 6.2

1. Write algebraic expressions for the (a) minterm and (b) maxterm realizations of the single-bit sum and carry functions.
2. We saw a comparator based on a circuit that cascaded left-to-right, that is, that started with the most significant bit and in which information was passed to bit positions of successively less significance until the end of the numbers being compared was reached. Design a cascaded comparator that compares two n-bit numbers and produces two output bits: one to indicate $A = B$ and one to indicate $A > B$, but which cascades from right to left. That is, information is passed within the comparator from the cell in the least significant bit position to bit positions of successively greater significance until the output of the cell in the most significant bit position is taken as the output of the entire comparator.

6.5 THE ALU

We can now apply our knowledge of adders, multiplexers, and multi-bit parallel operations to build a controllable, multiple-function digital circuit called an **arithmetic logic unit** (ALU). The ALU is the heart of a central processing unit (CPU), which itself is the heart of a computer. In a rather limited sense, it is a computer itself, in that it performs any one of a variety of functions depending on an externally provided command.

Our simple ALU will deal with six-bit numbers. It will yield a six-bit number as output, and it will use a three-bit control signal C to determine what operation to perform on its two six-bit data inputs, A and B. At a high level, then, our ALU is shown in Figure 6.16.

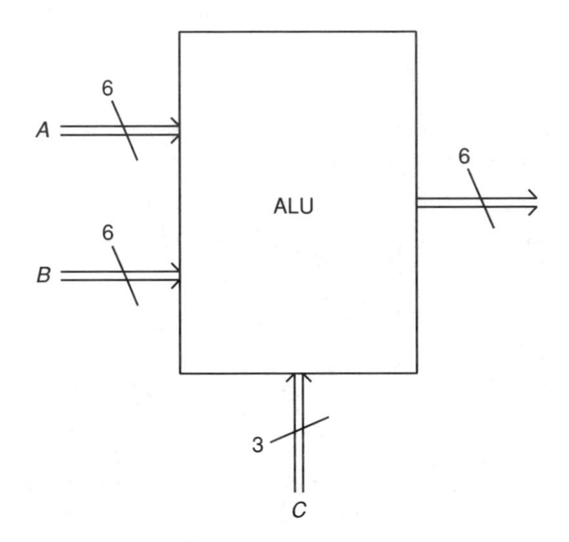

Figure 6.16 High-level block diagram of a simple ALU.

Because there are three bits in the control signal, there is a total of eight different commands that we can give the ALU. These commands range from 000_2 to 111_2 (0 through 7) and are called **opcodes** for *operation codes* because they command the ALU to perform particular operations. Our ALU is to perform its operations according to the chart in Table 6.5.

TABLE 6.5

Opcode (base 10)	Opcode (base 2)	ALU Output
0	0 0 0	0 0 0 0 0 0
1	0 0 1	A
2	0 1 0	B
3	0 1 1	$A + B$
4	1 0 0	$A \wedge B$
5	1 0 1	$A \vee B$
6	1 1 0	$A \oplus B$
7	1 1 1	1 1 1 1 1 1

This table is not exactly a truth table. It tells what function the ALU will perform for every possible value of its input opcode (0–7) but it does not tell the specific output for each possible value of A and B. Its output column is given in terms of A and B, not in terms of 0's and 1's.

The output of the ALU is a six-bit number. When the opcode is 0, the ALU's output should be all zeros. Likewise, when the opcode is 7, the ALU's output should be all ones. In these cases, the six-bit inputs A and B are ignored. If the opcode is 1, the ALU simply passes A through, yielding a six-bit output that is whatever A is. Similarly, if the opcode is 2, the ALU's output is whatever B is. If the opcode is 3, the ALU's output is the sum of A and B. If the opcode is 4, 5, or 6, the output is the result of the bitwise Boolean operations AND, OR, and XOR, respectively, performed on A and B.

Because the output from the ALU is six bits and A and B are also six bits each, it is possible that adding A and B together (opcode 3) could produce a number too big to fit in the output bits. Such a situation is called an **overflow.** For the purposes of simplicity we will not worry about overflows in our ALU.[1] If $A + B$ happens to yield a seven-bit number, the output of the ALU will just be the least significant six bits of the sum; the seventh (most significant) bit will be discarded. The desired behavior of the ALU is summarized by the following quasi-algebraic function:

$$
\begin{aligned}
\text{output} = {} & [0 \wedge (C = 0)] \vee \\
& [A \wedge (C = 1)] \vee \\
& [B \wedge (C = 2)] \vee \\
& [(A + B) \wedge (C = 3)] \vee \\
& [(A \wedge B) \wedge (C = 4)] \vee \\
& [(A \vee B) \wedge (C = 5)] \vee \\
& [(A \oplus B) \wedge (C = 6)] \vee \\
& [111111 \wedge (C = 7)].
\end{aligned}
$$

The fact that the output of the ALU varies as a function of a control signal should suggest a multiplexer. Indeed, at the core of our ALU is a three-control eight-data six-bit wide parallel multiplexer. We will connect the six-bit output vectors of other circuits to the eight six-bit data inputs of the multiplexer in such a way as to satisfy the requirements specified in Table 6.5.

Specifically, we hardwire MUX inputs 0 and 7 to be all 0's and all 1's, respectively, and connect inputs A and B directly to MUX inputs 1 and 2. We know how to build six-bit parallel AND, OR, and XOR functions to perform

[1] Although in real ALUs, overflows, as well as various other internal conditions, are handled by setting appropriate bits in a so-called **status register** that is accessible to other circuits that make use of the ALU.

bitwise operations on the input vectors, so we construct these functions, using *A* and *B* as input, and connect their six-bit outputs to MUX inputs 4, 5, and 6. Finally, we construct a six-bit cascading adder, feed it *A* and *B* as inputs, and connect the first six bits of its output to MUX input 3. The implementation of the ALU is shown in Figure 6.17.

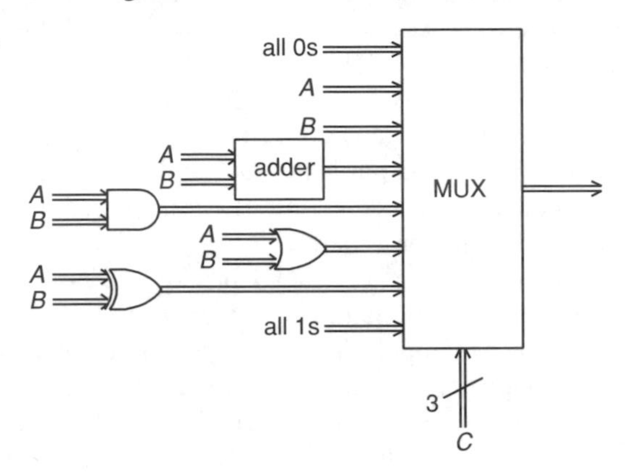

Figure 6.17 Block diagram of a six-bit ALU. All vectors are six bits wide, except the control vector C.

The examples given in this chapter represent a tiny cross section of the world of digital circuitry. Nonetheless, the principles used here are the same as those used in all digital circuits. There is one important conceptual piece that will not be covered here, but it is essential to the design of many of the more complex circuits. This piece is **memory.** Memory circuits allow us to store data for a circuit to manipulate, but more importantly to circuit design, they allow us to create **state machines.** These are circuits that yield different output signals based not only on externally provided input data and control signals, but also on the basis of an internally stored **state,** as manifested by internal memory circuitry. A state machine is responsible for not only generating the proper output data bits, but also for updating its own internal state according to its design specifications on the basis of the current state and the externally provided input signals. Since the behavior of a state machine depends on its state at any given time, the state allows it to "remember" where it is in the middle of a complex computation and behave differently based on this. Thus, state machines make possible circuits with a stored "program." For this reason, state machines are capable of carrying out a sequence of operations resulting in considerably more complex behavior than is possible with the types of circuits presented here.

EXERCISE 6.3

1. How difficult would it be to implement the six-bit ALU discussed in Section 6.5 simply by listing each of its individual input bits and output bits and deriving a minterm or

maxterm realization of it? Specifically, how many individual functions would we need to define? How many inputs would each function have? How many lines would there be in the truth tables for each of these functions?

Show how the following actual digital circuits might be implemented and draw block diagrams for each of the following:

2. An encoder: a circuit with n outputs and 2^n inputs. The output bits are taken together to form an n-bit binary number that reflects which of the input lines is asserted. For example, if input line 3 is asserted, the output should be (011), i.e., 3_{10}. Assume that only one of the input bits will ever be asserted at any time.
3. The demultiplexer: a circuit with one data input (x), an n-bit control signal, and 2^n one-bit output lines. The control determines which output line x goes out on. All other output lines are to be 0. Note that through bit slicing and parallelism, x and the outputs can be signals of any width.
4. Draw a block diagram of an eight-bit ALU that behaves according to the following chart. Note that there is only one eight-bit vector of data input, A, and only two bits in the control signal.

Opcode	ALU Output
00	A
01	$A \div 2$
10	$A \div 4$
11	$A \div 8$

The divisions are integer division; any remainder is to be ignored. (*Hint:* note that the divisions are by powers of two. How would we easily implement division of binary numbers by numbers that are powers of two? Remember how we divide base-10 numbers by powers of 10.)

7

Laws of Boolean Algebra

Although we now know how to create minterm or maxterm realizations of any Boolean function for which we can write a truth table, we have also seen that these realizations sometimes lead to unnecessarily complicated expressions, as in the case of the three-input function whose truth table is shown in Table 7.1.

The minterm realization of this function is $M_1 \vee M_3 \vee M_5 \vee M_7$. However, The output column matches the third input bit, z, exactly. The algebraic expression for the simplest possible realization of this function is, in fact, "z." In this chapter we will explore some of the algebraic rules that allow us to manipulate Boolean expressions, transforming and thereby simplifying them.

In the old days of digital circuit design, a single logic gate might have been constructed out of several vacuum tubes each costing tens of cents. Even small combinational logic networks could get prohibitively expensive and bulky with accompanying problems of cooling, increased failure rates, and the like. Thus, early designers of digital circuits were highly motivated to minimize and simplify their circuits whenever possible. While today we are freed from the clumsy hardware of the past, semiconductor chip designers are now under intense pressure to pack as much computational power onto a single chip of silicon as they can. This involves physical hardware advances (making the electrical realization of a single logic gate physically smaller, making them run cooler so more can be placed closer together without overheating, etc.) as well as algebraic manipulation to make a given logic function simply require fewer logic gates in the first place.

TABLE 7.1

x	y	z	
0	0	0	0
0	0	1	1
0	1	0	0
0	1	1	1
1	0	0	0
1	0	1	1
1	1	0	0
1	1	1	1

7.1 SETS OF AXIOMS

Boolean algebra is not one mathematical system, but rather an entire family of mathematical systems of which the one involving zeros and ones and AND, OR, and NOT is but one example (albeit the most common one, and the one most people think of when they hear the term *Boolean algebra*). The algebra of sets is another. Technically, any mathematical system is a Boolean algebra if it conforms to certain **axioms.** Axioms are the fundamental, unproven truths within a mathematical system from which the **theorems** are derived by explicit mathematical proof. They are the (preferably few) statements that are taken as given within the system, and upon which everything else in the system is built. Thus, the initial set of axioms characterizes the type of mathematical system built on them. If all of the axioms of Boolean algebra are true of a given mathematical system, then that system is a Boolean algebra. As we encounter the axioms of Boolean algebra in this chapter, we will show that each of them is true of our 0/1, AND/OR/NOT system.

Over the years many mathematicians have written different sets of axioms to define the class of systems known as Boolean algebras, but creating sets of axioms is a subtle art. When defining categories of systems, mathematicians mix and match axioms to come up with sets of axioms that conform to various criteria. First of all, the axioms must never, ever contradict each other. This may seem obvious, but sometimes implicit contradictions can lie buried only to manifest themselves after many theorems have been written. Sets of axioms containing contradictions are said to be **inconsistent.** Also, if a given axiom is found to be derivable from the other axioms then it is no longer considered an axiom; it is a theorem. As mathematicians like the smallest possible set of axioms that allow them to prove the greatest number of theorems, such redundancy is weeded out, although this too can be an elusive goal.

We will begin with some axioms and then prove some theorems as well, thereby building up a substantial body of formally derived knowledge about our system of Boolean algebra. As we establish rules of greater complexity and power, we will be rigorously explicit as to what we know and what we do not know, and how we prove new rules. Some of this process may be familiar to you from traditional algebra, and some of the rules we prove may even seem similar. But watch out! Some truths of Boolean algebra do not hold true when applied to "normal" numbers.

We will use truth tables as we go along to show that the Boolean algebra we have been studying so far does in fact agree perfectly with the axioms. Specifically, when we want to show that two seemingly different expressions are actually equivalent, we will write the truth tables for both expressions and compare them line by line. If they match exactly, the expressions are equivalent. Such a rule is called an **equality** and can be written as an **equation,** that is, two expressions side-by-side with an equals sign between them. Most of our rules will take this form.

Likewise, if we compare the truth tables of two expressions and they differ in each row, then we will have shown that the expressions are complementary, as when we showed that $x \oplus y$ is the complement of $x \odot y$. Either way, because we make sure that the rule we want to prove holds true in each line of the truth tables involved, we know that it is true for all possible cases: $x \oplus y = \overline{(x \odot y)}$ no matter what x and y are. This method of proof is known as **perfect induction.** By "proving" the axioms of Boolean algebra in terms of our 0 and 1, and AND, OR, and NOT, we will be verifying that the system we have been studying is, in fact, a Boolean algebra. Looked at from another, more practical point of view, we will be showing that the axioms themselves hold true when applied to AND, OR, and NOT as we have understood them so far.

7.2 PERFECT INDUCTION

Induction is the type of reasoning we use when we generalize on the basis of a limited number of specific examples. One might **induce,** for instance, that all birds fly after seeing many different kinds of birds fly. One would be incorrect in that case, of course, as one look at a penguin or ostrich will show you. This is why induction must be used carefully in real life. Generalizing from one's inherently incomplete experience can lead to incorrect conclusions.

In the world of mathematics and logic, however, we can prove things beyond a shadow of a doubt using perfect induction in which we generalize, not on the basis of a few cases or most cases, but on the basis of *all possible cases.* If we see that a rule is true for each particular row on a truth table, we may generalize with confidence that the rule is always true. Perfect induction is sometimes re-

TABLE 7.2

x	$0 \vee x$
0	0
1	1

ferred to as the "brute force" method, because when we use it we just pound through all the possibilities without any clever leaps of logic.[1]

The first axioms of Boolean algebra that we will learn are called the special properties of 0 and 1.

7.2.1. Special Properties of 0 and 1

$$0 \vee x = x \tag{7.1}$$

$$1 \wedge x = x \tag{7.2}$$

$$1 \vee x = 1 \tag{7.3}$$

$$0 \wedge x = 0 \tag{7.4}$$

These equations certainly agree with our understanding of AND and OR so far—anything ORed with 0 equals itself; anything ANDed with 0 equals 0; anything ORed with 1 equals 1; and anything ANDed with 1 equals itself. We can formally verify this agreement using perfect induction, in effect "proving" the axioms in terms of what we already know about our familiar functions.

Since each of the four special properties of 0 and 1 contains only one variable, the truth tables we use to prove them using perfect induction have only two rows. For now we will only treat the special properties of 0 and 1 that deal with OR (Table 7.2, 7.3).

Looking at Table 7.2, we see that when x is 0, the output is 0 as well, and that when x is 1, the output is also 1. Since the output column matches x in each

TABLE 7.3

x	$1 \vee x$
0	1
1	1

[1] ... 'enumeration of cases,' indeed, is one of the duller forms of mathematical argument." G. H. Hardy, *A Mathematician's Apology*, p. 113.

of the two rows, we have shown by perfect induction that $0 \vee x$ always equals x itself, no matter what value x has.

Since there are nothing but 1's in the output column of the truth table shown in Table 7.3, we have inductively shown that the function $1 \vee x$ is a tautology: $1 \vee x$ ignores its input, always yielding 1 as output.

Before we continue with more axioms, it should be pointed out that any Boolean algebraic expression ultimately represents a single bit, no matter how complicated the expression. For some combinations of inputs, an expression may be 0 and for some it may be 1, but it always boils down to a bit. Therefore, while we use simple single letter variable names to represent bits (like x, y, and z) when we prove our rules, in practice we may apply these rules by substituting entire complicated algebraic expressions in place of the variables.

For example, the special properties of 0 tell us not only that $0 \vee x = x$, but that $0 \vee [a \wedge (\bar{b} \oplus c)] = [a \wedge (\bar{b} \oplus c)]$. The idea is the same: any Boolean expression ORed with 0 equals itself. We may even substitute $\bar{x}$ for x in the special property of 0, thus showing that $0 \vee \bar{x} = \bar{x}$.

7.2.2 The Complementation Laws

The next axioms of Boolean algebra we need to know are called the complementation laws because they concern how AND and OR deal with bits and their complements.

$$x \vee \bar{x} = 1 \tag{7.5}$$

$$x \wedge \bar{x} = 0 \tag{7.6}$$

This pair of axioms tells us that any bit ORed with its complement equals 1, and any bit ANDed with its complement equals 0. No matter what value a bit has, if we AND or OR it with its complement, we will end up ANDing or ORing a 0 and a 1 together. Since according to our prior understanding of AND and OR, $0 \wedge 1 = 0$ and $0 \vee 1 = 1$, the complementation laws should come as no surprise, but we will verify them with perfect induction.

As equations 7.5 and 7.6 concern only one bit as input, the truth table needed to prove each of them need only be two lines long. For the sake of convenience, we will use the same truth table for both Table 7.4.

For clarity, the intermediate column representing the value of $\bar{x}$ for each of

TABLE 7.4

x	$\bar{x}$	$x \vee \bar{x}$	$x \wedge \bar{x}$
0	1	1	0
1	0	1	0

the two values of x is shown. As expected, in each of the two rows in Table 7.4 there is a 1 in the output column corresponding to $x \bigvee \bar{x}$ and a 0 in the column corresponding to $x \bigwedge \bar{x}$. Thus these functions ignore their inputs and are equal to constant 1 and 0 functions, respectively.

7.2.3 The Law of Involution

The next axiom of Boolean algebra is simple. It merely establishes that the NOT function cancels itself out:

$$\bar{\bar{x}} = x. \tag{7.7}$$

The complement of the complement of any bit equals that bit, as shown in the truth table in Table 7.5.

TABLE 7.5

x	$\bar{x}$	$\bar{\bar{x}}$
0	1	0
1	0	1

Again, there is only one variable bit involved, so the truth table has only two lines in it. The output column corresponding to $\bar{\bar{x}}$ matches the input column corresponding to x in each of the two rows, so $x = \bar{\bar{x}}$ for all possible values of x, and thus is always true.

7.2.4 Commutative Laws of AND and OR

This pair of laws states that AND and OR are symmetric functions; the output of AND and OR in no way depends on the order of the input bits:

$$x \bigvee y = y \bigvee x \tag{7.8}$$

$$x \bigwedge y = y \bigwedge x. \tag{7.9}$$

Just as a commuter is someone who shuttles back and forth to work each day, the commutative laws of AND and OR tell us that we may **commute** the input bits of AND and OR back and forth without affecting the results of the operations.

Since each of the two equations 7.8 and 7.9 has two distinct bits in it, the truth table needed to prove these laws by perfect induction must have four rows (Table 7.6).

In each of the four rows the output for $x \bigvee y$ is the same as the output for $y \bigvee x$, just as the output bit in each of the four rows under $x \bigwedge y$ exactly matches

TABLE 7.6

x	y	$x \vee y$	$y \vee x$	$x \wedge y$	$y \wedge x$
0	0	0	0	0	0
0	1	1	1	0	0
1	0	1	1	0	0
1	1	1	1	1	1

the corresponding bit under $y \wedge x$. Thus, the outputs of AND and OR are unaffected by the order of their inputs, regardless of the values of those inputs.

7.2.5 Distributive Laws of AND and OR

The laws of Boolean algebra we have introduced have seemed rather common-sensical, even obvious. The distributive laws of AND and OR are not quite as intuitively clear:

$$x \wedge (y \vee z) = (x \wedge y) \vee (x \wedge z) \tag{7.10}$$

$$x \vee (y \wedge z) = (x \vee y) \wedge (x \vee z). \tag{7.11}$$

To understand exactly what these laws say, it might be useful to illustrate equation (7.10), the distributive law of AND over OR, with a circuit diagram (see Fig. 7.1).

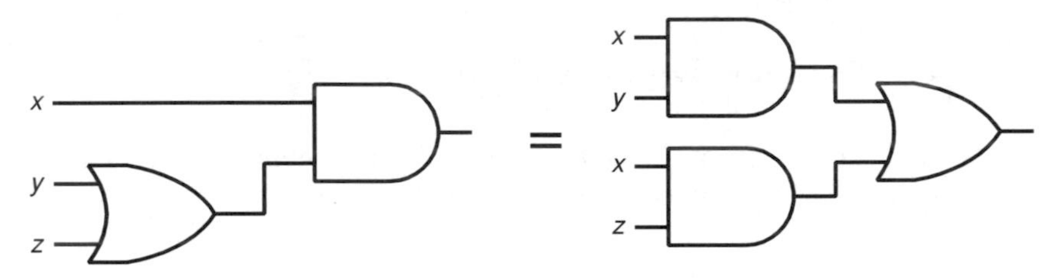

Figure 7.1 Circuit diagram illustrating the distributive law of AND over OR.

The laws are so named because when AND is applied to the result of a separate OR function for example, we can **distribute** the AND, applying it first to the inputs of the OR, then ORing at the end. In the algebraic expression, instead of ANDing x with the result of the OR inside the parentheses, we can AND the x with the bits inside the parentheses separately (y and z) and perform the OR afterwards on the two resulting bits.

To prove the distributive law of AND over OR by perfect induction, we need a three-input truth table (Table 7.7). Some of the intermediate columns have been shown in Table 7.7 to make the logic easier to follow. The highlighted

TABLE 7.7

x	y	z	$y \vee z$	$x \wedge (y \vee z)$	$x \wedge y$	$x \wedge z$	$(x \wedge y) \vee (x \wedge z)$
0	0	0	0	0	0	0	0
0	0	1	1	0	0	0	0
0	1	0	1	0	0	0	0
0	1	1	1	0	0	0	0
1	0	0	0	0	0	0	0
1	0	1	1	1	0	1	1
1	1	0	1	1	1	0	1
1	1	1	1	1	1	1	1

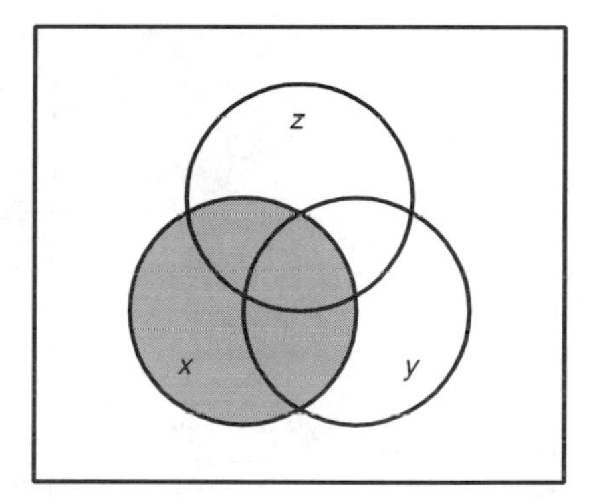

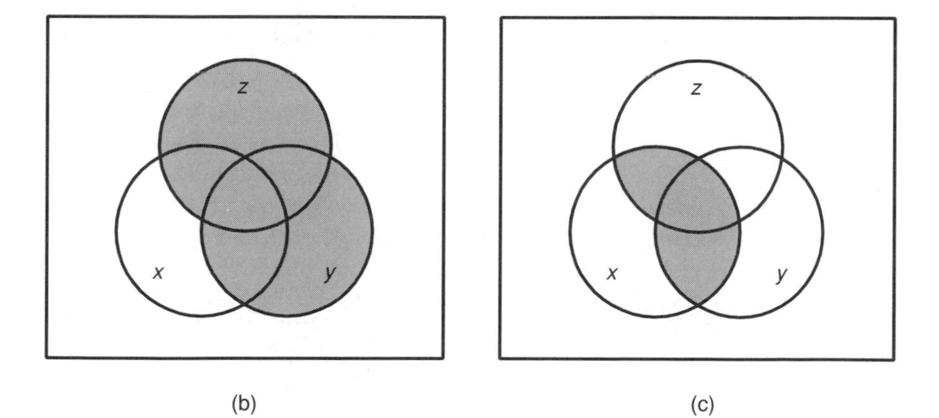

(b) (c)

Figure 7.2 Three-set Venn diagrams of (a) x, (b) $y \cup z$, (c) $x \cap (y \cup z)$.

columns represent the output columns of those two expressions that the distributive law of AND over OR asserts are equal. Because they are indeed equal for each of the eight possible combinations of values that the three input bits could assume, the distributive law of AND over OR is always true.

Sometimes it is easier to see the truth of algebraic truths by using a Venn diagram to illustrate the different steps in a proof by perfect induction. The distributive law of AND over OR has three variables, so we need a three-set Venn diagram. We will draw the Venn diagrams for both sides of the equation, and we should see that they are identical.

First, we will illustrate the first side of the equation by drawing the Venn diagrams for x and $(y \cup z)$ and their intersection as shown in Figure 7.2.

We can illustrate the right side of the equation by drawing the Venn diagrams for the sets $x \cap y$, $x \cap z$ and their union as shown in Figure 7.3. As expected, the two final Venn diagrams are identical, confirming our proof by perfect induction.

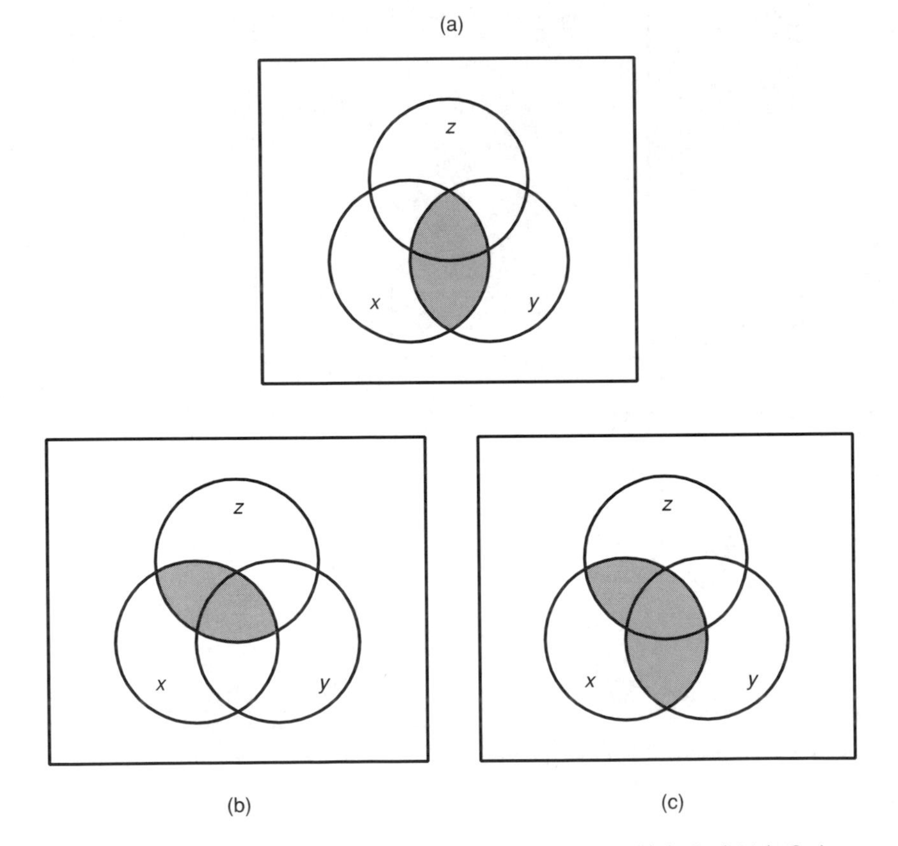

Figure 7.3 Three-set Venn diagrams for (a) $x \cap y$, (b) $x \cap z$, (c) $(x \cap y) \cup (x \cap z)$.

EXERCISE 7.1

A. Prove the following by perfect induction:

1. The special properties of 0 and 1 relating to AND (Equations 7.2 and 7.4).
2. The distributive law of OR over AND (Equation 7.11).

B. Which laws are illustrated by the following equations?

3. $x \vee \bar{x} = \bar{x} \vee x$
4. $(n \oplus m) \wedge 0 = 0$
5. $(a \vee \bar{b}) \vee (b \wedge c) = [(a \vee \bar{b}) \vee b] \wedge [(a \vee \bar{b}) \vee c)$
6. $\bar{\bar{\bar{x}}} = \bar{x}$
7. Draw circuit diagrams illustrating each of the rules presented so far, except the distributive law of AND over OR (equation 7.10), which has already been done.
8. Draw a Venn diagram illustrating the distributive law of OR over AND (equation 7.11).

7.3 DEDUCTION

If induction allows us to generalize from particular cases to general rules (and in the case of perfect induction, from all possible particular cases), deductive reasoning works in the other direction, allowing us to apply general rules to particular cases. The classic example of deduction is that if we already know that all men are mortal, and that Socrates is a man, then we may **deduce** that Socrates is mortal. This particular deduction is an example of a **syllogism,** a form of deduction we will look at more closely later. Deduction allows us to use the axioms and theorems that we have verified within our system to prove new theorems.

Generally, when we prove a Boolean equality using deduction, we begin with one Boolean expression and transform it step by step into different but equivalent expressions using previously proven laws until it appears in the form of the desired equivalent expression.

7.4 ALLOWED MANIPULATIONS OF BOOLEAN EQUATIONS

Before we continue, we should establish the types of algebraic operations we are allowed and not allowed to perform on Boolean equations.

Once we have proven a rule, its further utility to us lies in the fact that we may "plug in" or substitute any Boolean expression we want to for the variables in that rule and still have a true statement. An expression we substitute for the variables in a rule may consist of different variables, complemented or uncomplemented, or even long parenthetical expressions. However, we must be aware of the limits of any rule we want to apply. One such limitation is that if a variable appears more than once in the formulation of a rule, the same expression must be substituted for each appearance of that variable.

For example, one of the Special Properties of 1 tells us that $1 \wedge x = x$. We may substitute the expression $[(x \wedge y) \vee z]$ for the variable x in this rule, as long as we substitute this exact expression for each occurrence of x in the formulation of the rule. Thus the equation

$$1 \wedge [(x \wedge y) \vee z] = [(x \wedge y) \vee z]$$

is deductively proven to be true by application of the Special Property of 1.

Likewise, if a rule proves some property of two expressions ANDed together, it may not be applied to a situation in which three expressions are ANDed together without first proving that the property holds true in the three-expression case. We begin by proving rules for the two-input AND and OR function case, then explicitly expand them using other rules.

If we have a Boolean equation that we know to be valid (i.e., both sides of the equation are, in fact, equal to each other), we are allowed to perform any operation we want on *both* sides of the equation, and we will still have a valid equation. We may complement both sides of the equation, AND the same (or different, but equivalent) expressions onto both sides, or OR the same (or different but equivalent) expressions onto both sides.

For example, the complementation law of OR (equation 7.5) tells us that $x \vee \bar{x} = 1$. We may generate a perfectly true, but different equation by ANDing both sides with the expression $(x \wedge z)$ to produce

$$(x \vee \bar{x}) \wedge (x \wedge z) = 1 \wedge (x \wedge z).$$

Likewise, because we know that $x \vee 0$ and x are equivalent expressions (from the Special Property of 0, equation 7.1), we may OR each of these with the corresponding side of the complementation law of OR to give us the valid equation

$$(x \vee \bar{x}) \vee (x \vee 0) = 1 \vee x.$$

7.4.1 Idempotence

The first theorem we shall prove using a deductive proof is one of the two laws of idempotence:

$$x \vee x = x \tag{7.12}$$

$$x \wedge x = x. \tag{7.13}$$

The word *idempotence* comes from the Latin *idem,* meaning the same, and *potent,* meaning power. In mathematical terms, idempotence is the property a quantity has if the product of itself multiplied by itself (raised to the power of 2) equals itself. In the realm of ordinary, so-called natural numbers, only 0 and 1 are idempotent ($0^2 = 0$ and $1^2 = 1$). The application of the word *idempotent* to a theorem in Boolean algebra is another instance of the analogy between the arithmetic function of multiplication and the Boolean function AND (and here, OR as well).

To prove equation 7.12 deductively, we write down the portion on the left-hand side of the equals sign:

$$x \vee x.$$

We will write successive transformations of the original expression, each on a separate line, until we have the desired final expression, in this case, x. We will also write an equals sign after each new form of the expression to make the continuity of equality more explicit. To the right of each new form of the expression, we will write the name of the law that justifies that particular transformation. We will begin by using one of the Special Properties of 1, equation 7.2:

$$\begin{aligned} x \vee x &= \\ 1 \wedge (x \vee x) & \qquad (7.2). \end{aligned}$$

Equation 7.2 states that anything ANDed with 1 equals itself. We have used that law to AND the expression in the first line, $x \wedge x$, with 1 to produce the equivalent expression beneath it. In effect, we have substituted the expression $(x \wedge x)$ for the variable x in equation 7.2.

Because equation 7.5 (the complementation law of OR) states that $x \vee \bar{x} = 1$, we may turn the 1 in the resulting expression into $x \vee \bar{x}$ without changing the value of the expression as a whole.

$$\begin{aligned} x \vee x &= \\ 1 \wedge (x \vee x) &= \qquad (7.2) \\ (x \vee \bar{x}) \wedge (x \vee x) &= \qquad (7.5). \end{aligned}$$

The last expression we produced has two parenthetical expressions ANDed together with each containing x ORed with something else. This is the same form as the expression on the left-hand side of the equals sign in equation 7.11 (the distributive law of OR over AND). This means that we may now transform this expression into the form on the right hand side of the equals sign in equation 7.11, in effect, undistributing the OR from each of the two parenthetical expressions:

$$\begin{aligned} x \vee x &= \\ 1 \wedge (x \vee x) &= \qquad (7.2) \\ (x \vee \bar{x}) \wedge (x \vee x) &= \qquad (7.5) \\ x \vee (\bar{x} \wedge x) &= \qquad (7.11). \end{aligned}$$

Now we should turn the part of the equation still in parentheses into a 0 according to equation 7.6 (the complementation law of AND), which tells us that anything ANDed with its complement equals 0. However, the law states that $x \wedge \bar{x} = 0$, not that $\bar{x} \wedge x = 0$. The idea, of course, is the same, but we should be careful about form. We could simply use the commutative law of AND to switch the order of the variables inside the parentheses before applying the complementation law, but instead we will do something a little trickier. We will use the law of involution to turn the x into $\bar{\bar{x}}$

$$\begin{array}{rl}
x \vee x = & \\
1 \wedge (x \vee x) = & (7.2) \\
(x \vee \bar{x}) \wedge (x \vee x) = & (7.5) \\
x \vee (\bar{x} \wedge x) = & (7.11) \\
x \vee (\bar{x} \wedge \bar{\bar{x}}) = & (7.7).
\end{array}$$

Now we may apply equation 7.6 because inside the parentheses we have an expression ANDed with its complement in which the uncomplemented expression appears first. In this case, the expression is $\bar{x}$ and its complement is $\bar{\bar{x}}$. We are really substituting $\bar{x}$ for x in equation 7.6:

$$\begin{array}{rl}
x \vee x = & \\
1 \wedge (x \vee x) = & (7.2) \\
(x \vee \bar{x}) \wedge (x \vee x) = & (7.5) \\
x \vee (\bar{x} \wedge x) = & (7.11) \\
x \vee (\bar{x} \vee \bar{\bar{x}}) = & (7.7) \\
x \vee 0 = & (7.6).
\end{array}$$

We know from equation 7.1 that anything ORed with 0 equals itself, so we should be able to get rid of the 0 in the last expression, but equation 7.1 states that $0 \vee x = x$, not, as we have, that $x \vee 0 = x$. So we first apply the commutative law of OR to get the parts of the expression in the proper order, then use the special property of 0 to dispose of the 0:

$$\begin{array}{rl}
x \vee x = & \\
1 \wedge (x \vee x) = & (7.2) \\
(x \wedge \bar{x}) \wedge (x \vee x) = & (7.5) \\
x \vee (\bar{x} \wedge x) = & (7.11) \\
x \vee (\bar{x} \wedge \bar{\bar{x}}) = & (7.7)
\end{array}$$

$$x \vee 0 = \qquad (7.6)$$

$$0 \vee x = \qquad (7.8)$$

$$x \qquad (7.1)$$

Q.E.D.

The final x on the right of the quals sign tells us that we are done, we have proven deductively that $x \vee x = x$. The letters Q.E.D. are an acronym for the latin expression *quod erat demonstrandum,* literally "that which was to be demonstrated." These letters traditionally signal the end of a formal deductive proof of the type we just created. It is sort of the scholarly equivalent of "voila!" or "And there you have it!"

Following then is the proof of equation 7.13:

$$x \wedge x =$$

$$0 \vee (x \wedge x) = \qquad (7.1)$$

$$(x \wedge \bar{x}) \vee (x \wedge x) = \qquad (7.6)$$

$$x \wedge (\bar{x} \vee x) = \qquad (7.10)$$

$$x \wedge (\bar{x} \vee \bar{\bar{x}}) = \qquad (7.7)$$

$$x \wedge 1 = \qquad (7.5)$$

$$1 \wedge x = \qquad (7.9)$$

$$x \qquad (7.2)$$

Q.E.D.

7.4.2 Absorption Laws

We will now look at six theorems grouped under the heading **absorption laws** because they permit shrinking of Boolean expressions by **absorbing** some terms into other terms. However, we will just as often use them as "expansion laws" to turn simple expressions into more complex ones. The absorption laws will be used often in the material ahead and are powerful, although not intuitively obvious.

$$x \vee (x \wedge y) = x \qquad (7.14)$$

$$x \wedge (x \vee y) = x \qquad (7.15)$$

$$x \vee (\bar{x} \wedge y) = x \vee y \qquad (7.16)$$

$$x \wedge (\bar{x} \vee y) = x \wedge y \qquad (7.17)$$

$$(x \wedge y) \vee (x \wedge \bar{y}) = x \qquad (7.18)$$

$$(x \vee y) \wedge (x \vee \bar{y}) = x \qquad (7.19).$$

A deductive proof of equation 7.14 follows:

$$\begin{aligned} x \vee (x \wedge y) &= \\ (1 \wedge x) \vee (x \wedge y) &= && (7.2) \\ (x \wedge 1) \vee (x \wedge y) &= && (7.9) \\ x \wedge (1 \vee y) &= && (7.10) \\ x \wedge 1 &= && (7.3) \\ 1 \wedge x &= && (7.9) \\ x & && (7.2) \end{aligned}$$

Q.E.D.

Note that in the first step we made the expression temporarily more complicated to get it into a form to which we could apply the distributive law. The proof of equation 7.15 is a bit shorter:

$$\begin{aligned} x \vee (\bar{x} \wedge y) &= \\ (x \vee \bar{x}) \wedge (x \vee y) &= && (7.11) \\ 1 \wedge (x \vee y) &= && (7.5) \\ x \vee y & && (7.2) \end{aligned}$$

Q.E.D.

In the first step we undistributed the OR, lengthening the expression so that we could apply the complementation law in the next step.

The proofs of the other four absorption laws are left as exercises for the reader.

7.4.3 Associativity Laws

Whereas the commutativity laws told us that we can commute, or interchange, expressions that we AND and OR together without changing their value, the associativity laws tell us that we may **associate** different expressions that we AND and OR together by using parentheses without changing their value.

$$(x \vee y) \vee z = x \vee (y \vee z) \qquad (7.20)$$

$$(x \wedge y) \wedge z = x \wedge (y \wedge z). \qquad (7.21)$$

The deductive proof of equation 7.21 is a bit tricky:

$$
\begin{aligned}
(x \wedge y) \wedge z &= & \\
[(x \wedge y) \wedge z] \vee 0 &= & (7.1) \\
[(x \wedge y) \wedge z] \vee (x \wedge \bar{x}) &= & (7.6) \\
[((x \wedge y) \wedge z) \vee x] \wedge [((x \wedge y) \wedge z) \vee \bar{x}) &= & (7.11) \\
[x \vee ((x \wedge y) \wedge z)] \wedge [\bar{x} \vee ((x \wedge y) \wedge z)] &= & (7.8) \\
[(x \vee (x \wedge y)) \wedge (x \vee z)] \wedge [(\bar{x} \vee (x \wedge y)) \wedge (\bar{x} \vee z)] &= & (7.11) \\
[x \wedge (x \vee z)] \wedge [(\bar{x} \vee (x \wedge y)) \wedge (\bar{x} \vee z)] &= & (7.14) \\
[x \wedge (x \vee z)] \wedge [((\bar{x} \vee x) \wedge (\bar{x} \vee y)) \wedge (\bar{x} \vee z)] &= & (7.11) \\
[x \wedge (x \vee z)] \wedge [(1 \wedge (\bar{x} \vee y)) \wedge (\bar{x} \vee z)] &= & (7.5) \\
[x \wedge (x \vee z)] \wedge [(\bar{x} \vee y) \wedge (\bar{x} \vee z)] &= & (7.2) \\
x \wedge [(\bar{x} \vee y) \wedge (\bar{x} \vee z)] &= & (7.15) \\
x \wedge [\bar{x} \vee (y \wedge z)] &= & (7.11) \\
x \wedge (y \wedge z) & & (7.17)
\end{aligned}
$$

Q.E.D.

Essentially, to prove the associativity law of AND, we used the distributive law a number of times to expand the expression $(x \wedge y) \wedge z$ out to a complicated form, then used the absorption laws to boil it back down again the way we wanted it.

These laws tell us that any grouping of three expressions ANDed together or ORed together is equivalent to any other grouping. We originally defined the arbitray-input AND and OR gates to be equal to a particular grouping of inputs operated on by two-input AND and OR. The associativity laws now assure us that we can place parentheses any way we want when we have three expressions ANDed and ORed together without affecting the value of the function. Just as we showed that building a three-input AND function out of the old two-input kind gave us the tools to build a four-, five-, or thousand-input AND function, the associativity laws can be similarly cascaded upward to prove that all possible groupings of a given set of any number of variables ANDed together (or ORed together) are equivalent. It should be emphasized that these laws do not apply if ANDs and ORs are mixed together in an expression; they only apply to chains of variables or expressions ANDed together or chains of variables or expressions ORed together.

7.4.4 DeMorgan's Laws

Augustus DeMorgan (1806–1871) was a good friend of George Boole's, and, in fact, developed some of the same ideas independently of Boole. Although his work preceded Boole's, his system was designed to represent traditional Aristotelian logic using symbols. Boole's system, however, was intended to comprise a completely formulated logic system that stood on its own axioms and operated according to its own laws. Nonetheless, Boole felt indebted enough to DeMorgan to acknowledge him in his work, and the following laws still bear his name:

$$\overline{x \vee y} = \bar{x} \wedge \bar{y} \tag{7.22}$$

$$\overline{x \wedge y} = \bar{x} \vee \bar{y}. \tag{7.23}$$

These laws are powerful indeed and are by no means intuitively obvious. To illustrate exactly what they mean, some circuit diagrams are in order as shown in Figure 7.4. Proof of DeMorgan's laws by perfect induction are left as an exercise for the reader.

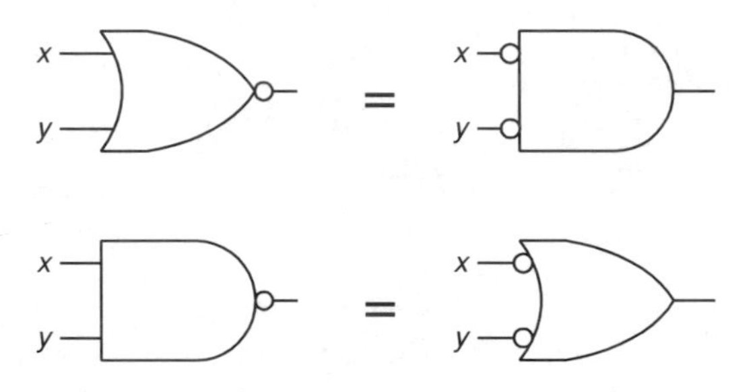

Figure 7.4 Circuit diagrams illustrating DeMorgan's laws.

The two DeMorgan laws give us new realizations of NAND and NOR. The left-hand side of each of the equations 7.22 and 7.23 are the definitions that we have been using for NOR and NAND, respectively, based on our understanding of them as NOT OR and NOT AND. The right hand side of the equations are the minterm realization of NOR and the maxterm realization of NAND, respectively.

An interesting consequence of DeMorgan's laws is that we can easily create an OR out of a single AND and some NOTs:

$$
\begin{aligned}
x \vee y &= \\
\overline{((\overline{x \vee y}))} &= \qquad (7.7) \\
\overline{(\bar{x} \wedge \bar{y})} &= \qquad (7.22).
\end{aligned}
$$

This is illustrated in the circuit diagram shown in Figure 7.5.

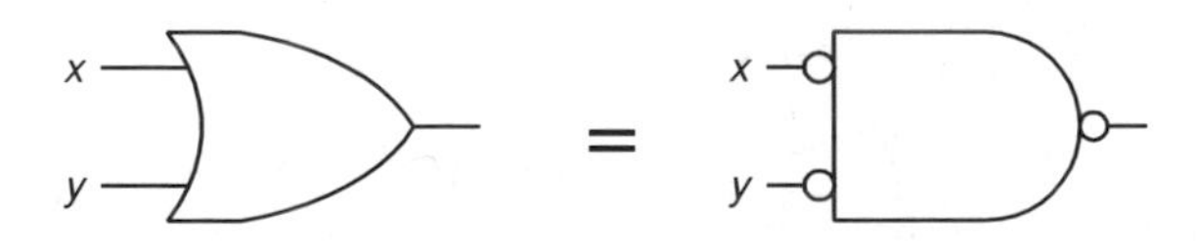

Figure 7.5 Realization of OR in terms of AND and NOT by way of DeMorgan.

We could also have arrived at this truth by substituting the values $\bar{x}$ and $\bar{y}$ for x and y, respectively in equation 7.23. Likewise, because two NOTs cancel each other out according to the law of involution, DeMorgan allows us to perform the transformation illustrated by the circuit diagrams in Figure 7.6.

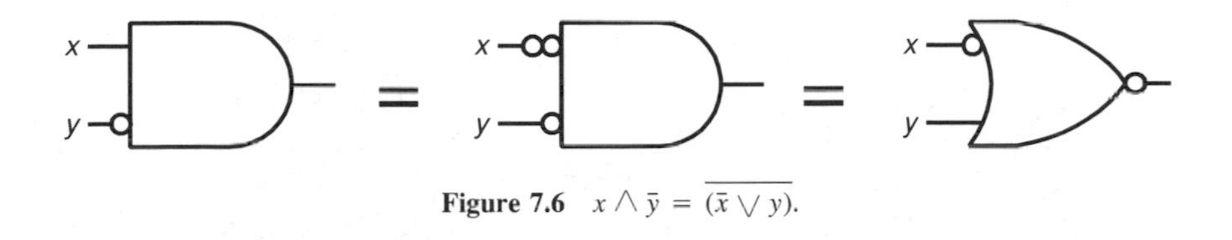

Figure 7.6 $x \wedge \bar{y} = \overline{(\bar{x} \vee y)}$.

As with the previous example, we might have achieved the same result by substituting $\bar{x}$ for x in equation 7.22. Indeed, any Boolean expression involving just one AND and some number of NOTs, either on the input or output of the AND, can be transformed by DeMorgan's laws into an equivalent expression involving one OR and some number of NOTs. For example, consider the pairs of equivalent expressions illustrated in Figure 7.7 (the middle one is the implication function).

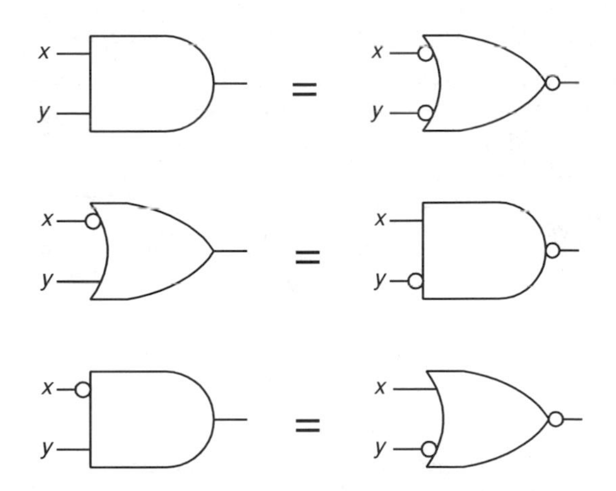

Figure 7.7 Allowed equivalences under DeMorgan's laws.

We now return to the troublesome example in chapter 6, in which the minterm realization of the two-input OR function was the unwieldy expression $(x \wedge \bar{y}) \vee (\bar{x} \wedge y) \vee (x \wedge y)$. We now have the tools to prove that this is equal to $x \vee y$:

$$
\begin{aligned}
(x \wedge \bar{y}) \vee (\bar{x} \wedge y) \vee (x \wedge y) &= \\
(x \wedge \bar{y}) \vee [(\bar{x} \wedge y) \vee (x \wedge y)] &= && (7.20) \\
(x \wedge \bar{y}) \vee [(y \wedge \bar{x}) \vee (y \wedge x)] &= && (7.9) \\
(x \wedge \bar{y}) \vee [(y \wedge (\bar{x} \vee x)] &= && (7.10) \\
(x \wedge \bar{y}) \vee [y \wedge (x \vee \bar{x})] &= && (7.8) \\
(x \wedge \bar{y}) \vee (y \wedge 1)] &= && (7.5) \\
(x \wedge \bar{y}) \vee (1 \wedge y) &= && (7.9) \\
(x \wedge \bar{y}) \vee y &= && (7.2) \\
y \vee (x \wedge \bar{y}) &= && (7.8) \\
y \vee (\bar{y} \wedge x) &= && (7.9) \\
y \vee x &= && (7.16) \\
x \vee y &= && (7.8)
\end{aligned}
$$

Q.E.D.

As another example, recall that the exclusive OR function is the same as the inclusive OR function except when x and y both equal 1. In this case, inclusive OR is 1, while exclusive OR is 0. Thus XOR is the same as IOR with the provision that both inputs are not 1. Articulated in Boolean terms, the exclusive OR function is $(x \vee y) \wedge \overline{(x \wedge y)}$; that is, x OR y, as long as NOT (x AND y). DeMorgan allows us to evaluate this:

$$(x \vee y) \wedge \overline{(x \wedge y)} = (x \vee y) \wedge (\bar{x} \vee \bar{y})$$

which, in fact, is the maxterm realization of the exclusive OR function that we derived in chapter 4.

EXERCISE 7.2

For a deductive proof in these exercises, the commutativity law need not be explicitly applied as a separate step.

A. Prove the following using perfect induction:

1. Both idempotence laws.
2. Both associativity laws.
3. The absorption laws.
4. For each of the absorption laws, draw a Venn diagram of the Boolean expression on each side of the equation showing any relevant intermediate expressions in separate Venn diagrams.

B. Illustrate the following axioms by drawing circuit diagrams of the expressions on both sides of the equals sign:

5. Both idempotence laws.
6. All six absorption laws.
7. Both associativity laws.

C. Prove the following deductively:

8. The idempotent law of AND (equation 7.13).
9. The absorption laws (equation 7.15 through 7.19).
10. The associativity law of OR (equation 7.20).
11. The maxterm realization of AND equals AND, i.e., $(x \vee y) \wedge (x \vee \bar{y}) \wedge (\bar{x} \vee y) = x \wedge y$.
12. The minterm realization of the implication function equals the implication function, i.e., $(\bar{x} \wedge \bar{y}) \vee (\bar{x} \wedge y) \vee (x \wedge y) = \bar{x} \vee y$.
13. The minterm realization of COIN equals the maxterm realization of COIN.

D. Recall that when we introduced the complete system based on NAND alone, we showed inductively that AND, OR, and NOT could be built from just NAND. Prove deductively that we can construct AND, OR, and NOT entirely out of NANDs. To apply the rules we learned above, take NAND to mean NOT AND in the following:

14. NOT constructed out of NAND: $\overline{(x \wedge x)} = \bar{x}$.
15. OR constructed out of NAND: $\overline{[\overline{(x \wedge x)} \wedge \overline{(y \wedge y)}]} = x \vee y$.
16. AND constructed out of NAND: $\overline{[\overline{(x \wedge y)} \wedge \overline{(x \wedge y)}]} = x \wedge y$.

E. Although we only proved that DeMorgan's laws and the distributivity laws are valid in the case of AND and OR functions with only two inputs, they may be extended to any number of variables we choose by using the extended associativity laws discussed in Section 7.4.3. As examples of how this might be done, prove the following by inserting parentheses as appropriate so that you can apply the two-input DeMorgan and distributivity laws.

17. Four-input DeMorgan: $\overline{(a \wedge b \wedge c \wedge d)} = \bar{a} \vee \bar{b} \vee \bar{c} \vee \bar{d}$.
18. Four-input distributivity: $a \wedge (b \vee c \vee d) = ab \vee ac \vee ad$.

F. As much as you can, simplify the following minterm or maxterm expressions using the algebraic laws:

19. The example from the beginning of the chapter, i.e., the function of three input bits whose maxterm realization is $M_1 \vee M_3 \vee M_5 \vee M_7$.
20. The minterm realization of the sum function from the adder circuit introduced in the last chapter, i.e., $M_1 \vee M_2 \vee M_4 \vee M_7$.

G. Convert the following expressions to minterm/maxterm form: i.e., blow them up. Do not cheat by using truth tables. (*Hint:* this will involve use of the absorption laws equations [equation 7.18 and 7.19].)

21. Maxterm realization of $z \wedge (\bar{x} \vee y)$.
22. Minterm realization of $\overline{(x \vee y)} \vee y$.
23. Maxterm realization of $\overline{(x \vee y)} \vee y$.

7.5 PRINCIPLE OF DUALITY

Once we have proven that a Boolean equation is valid we may derive an equally true second equation from the first by performing a simple two-step operation on the equation:

1. Turn all ANDs into ORs and vice versa.
2. Turn all explicit 0's into 1's and vice versa.

An equation derived in this way from another is called the **dual** of the original equation, and the fact that we may apply this technique to arrive at valid truths is known as the **principle of duality.** For example, assume that we have proven equation 7.2 by perfect induction. To find the dual of this rule, we first turn the AND into an OR:

$$1 \wedge x = x \qquad \text{becomes} \qquad 1 \vee x = x.$$

We then turn the 1 into a 0:

$$1 \vee x = x \qquad \text{becomes} \qquad 0 \vee x = x.$$

The equation we end up with, $0 \vee x = x$, should be recognizable as equation 7.1, which we already know to be true. The principle of duality allows us to prove two laws for the price of one. Having proven one Boolean equation, we automatically know that its dual is true as well without having to go through the process of a deductive or inductive proof.

It must be emphasized, however, that the principle of duality *does not* state that a Boolean expression is equal to its own dual; only that if two Boolean expressions are equal to each other, then their duals will also be equal to each other.

How can we always be sure that the dual of a proven equation will also be true? Note that all the equations we have proven so far have been presented in pairs—a law or set of laws on the left and its or their dual(s) on the right (the law of involution is its own dual; there are no ANDs, ORs, 0's or 1's in it). For example, the six laws of absorption are really three laws on the left and their duals on the right.

We know that the principle of duality holds true for all known laws so far because we have proven the duals of each law as we went along, either through perfect induction or deductive proof. Because the principle of duality holds true for all of our laws, it will also hold true for any new rules we prove using these laws. Whenever we construct a deductive proof of any new law, the deductive proof of its dual is exactly the same except that at each step we apply the dual of the law we applied in the corresponding step of the original proof. Thus, the proof of any law and the proof of its dual are almost mirror images of each other.

Take, for example, the proof of the absorption law given in equation 7.14 and its dual, equation 7.15.

Equation 7.14:

$$
\begin{aligned}
x \vee (x \wedge y) &= \\
(1 \wedge x) \vee (x \wedge y) &= && (7.2)\\
(x \wedge 1) \vee (x \wedge y) &= && (7.9)\\
x \wedge (1 \vee y) &= && (7.10)\\
x \wedge 1 &= && (7.3)\\
1 \wedge x &= && (7.9)\\
x & && (7.2)
\end{aligned}
$$

Q.E.D.

Equation 7.15:

$$
\begin{aligned}
x \vee (x \wedge y) &= \\
(0 \vee x) \wedge (x \vee y) &= && (7.1)\\
(x \vee 0) \wedge (x \vee y) &= && (7.8)\\
x \vee (0 \wedge y) &= && (7.11)\\
x \vee 0 &= && (7.4)\\
0 \vee x &= && (7.8)\\
x & && (7.1)
\end{aligned}
$$

Q.E.D.

Alternatively, having proven equation 7.14, we may derive its dual (equation 7.15) according to the following logic. If we substitute $\bar{x}$ and $\bar{y}$ for x and y in equation 7.14, we have the following equality:

$$
\begin{aligned}
\bar{x} &= \\
\bar{x} \vee (\bar{x} \wedge \bar{y}) &= && (7.14)\\
\bar{x} \vee \overline{(x \vee y)} &= && (7.22)\\
\overline{[x \wedge (x \vee y)]} &= && (7.23).
\end{aligned}
$$

Having reached this point, we may apply the NOT function to both sides of the equation, giving us:

$$\bar{\bar{x}} = \overline{\overline{[x \wedge (x \vee y)]}}$$

The involution rule applied to both sides of the equation gives us the desired

$$x = x \wedge (x \vee y).$$

Any proven Boolean equation may be converted into its dual by some sequence of applications of DeMorgan's law, the involution law, and the application of the NOT function to both sides of the equation.

The principle of duality would not hold true if we had used George Boole's original OR function, which was undefined in the case in which both inputs are 1. The desire to allow the principle of duality to work was a strong motivating force in the minds of Boole's followers (particularly Charles Sanders Peirce and Ernst Schröder) who redefined the OR function to be the inclusive OR that we use today rather than the variation of the exclusive OR function that Boole used.

As a point of interest, note that the maxterm realization of AND and the minterm realization of OR are duals of each other. That is

$$(x \vee y) \wedge (x \vee \bar{y}) \wedge (\bar{x} \vee y) = x \wedge y$$

is the dual of

$$(x \wedge y) \vee (x \wedge \bar{y}) \vee (\bar{x} \wedge y) = x \vee y.$$

EXERCISE 7.3

A. Write the duals of the three truths that prove that NAND is a complete system to show that NOR is a complete system as well:

1. NOT constructed out of NAND: $\overline{(x \wedge x)} = \bar{x}$.
2. OR constructed out of NAND: $\overline{[\overline{(x \wedge x)} \wedge \overline{(y \wedge y)}]} = x \vee y$.
3. AND constructed out of NAND: $\overline{(\overline{(x \wedge y)} \wedge \overline{(x \wedge y)})} = x \wedge y$.

8

Boolean Logic

George Boole and his contemporaries were not very interested in bits as we know them, and they certainly were not interested in adder circuits and multiplexers. However, they were quite interested in logical propositions and the algebra of sets. Specifically, Boole aimed to reduce all of human intelligence to more or less mechanical calculation through his system of symbolic logic. This involved putting a set of logical propositions, or statements, in symbolic form and manipulating the symbols algebraically to derive logical conclusions from them. We will now explore this application of Boolean algebra, which will provide a method for simplifying digital circuits as well as uncovering logical truths. Many of the ideas in the chapter were developed by Boole himself, but his work along these lines was extended by A. Blake in his 1937 dissertation, ''Canonical Expressions in Boolean Algebra.''

Before we begin, however, we must introduce some terminology.

8.1 OPPOSITION

In this chapter we shall work quite a bit with expressions in sum-of-products form. For this reason, unless otherwise noted, when we refer to a *term* we mean a product: one or more variables, complemented or not, ANDed together. We will often omit the AND symbol from our algebraic terms, writing $x\,\bar{y}\,\bar{z}$, for example, instead of $x \wedge \bar{y} \wedge \bar{z}$.

Two terms have an **opposition** if one contains a particular variable and the other contains the complement of that variable. For example, the terms $a\bar{b}c$ and $b\bar{d}$ have an opposition: the first contains $\bar{b}$ while the second contains b. Likewise, the terms $a\bar{b}c$ and $b\bar{c}d\bar{e}$ have two oppositions, b and c. ANDing together any two terms that have an opposition results in 0.

To prove this, let use the terms ax and $b\bar{x}$ to represent two terms that have an opposition between them. Clearly, they do have an opposition (x), and because we make no claims whatsoever about a and b (the remainder of either term except x and $\bar{x}$, respectively), this representation is sufficient to prove general truths about terms that have oppositions.

$$
\begin{aligned}
(a \wedge x) \wedge (b \wedge \bar{x}) &= \\
(a \wedge x) \wedge (\bar{x} \wedge b) &= && (7.9) \\
a \wedge (x \wedge \bar{x}) \wedge b &= && (7.21) \\
a \wedge 0 \wedge b &= && (7.6) \\
0 & && (7.4)
\end{aligned}
$$

Q.E.D.

8.2 CONSENSUS

When two people argue but come to an agreement, we say that they have reached a **consensus.** A consensus in this case is a compromise. The arguers put aside the things about which they disagreed and concentrated on those things on which they agreed (or at least did not disagree). Similarly, in the world of Boolean algebra if two terms have an opposition, their consensus is formed by ANDing the two terms together without the variable they have in opposition. For example, the consensus of the terms $a\bar{b}c$ and $b\bar{d}$ is $ac\bar{d}$. The terms were merged, but without the variable that they had in opposition, b.

If two terms have no opposition or have more than one opposition, they have no consensus. Thus, the terms $a\bar{b}c$ and $b\bar{c}\,\bar{e}$ have no consensus because they have two oppositions (b and c), nor do the terms $a\bar{b}c$ and $\bar{b}d\bar{e}$ that have no opposition. However, the terms $xy\bar{z}$ and $\bar{x}y\bar{z}$ have exactly one opposition, the variable x. They happen to contain exactly the same variables except the x, but they "disagree" on one and only one variable. The consensus is $y\bar{z}y\bar{z}$ or simplified, $y\bar{z}$. If x and y are terms, we denote the consensus of x and y as $c(x, y)$.

Our chief interest in the Boolean consensus lies in the following property. Given two terms ORed together in an SOP expression, ORing them with their consensus (if it exists) does not change the value of the expression. That is, given two terms x and y, $x \vee y = x \vee y \vee c(x, y)$.

To prove this, we assume that the two terms we are dealing with have an opposition. Therefore, we will once again represent our two terms with the expressions ax and $b\bar{x}$. The consensus of ax and $b\bar{x}$ is ab. What we would like to prove is that for any two terms that have an opposition, the following is true:

$$ax \lor b\bar{x} = ax \lor b\bar{x} \lor ab \tag{8.1}$$

$$\begin{aligned}
ax \lor b\bar{x} &= \\
(ax \lor axb) \lor (b\bar{x} \lor b\bar{x}a) &= \quad \text{(7.14) on each term} \\
(ax \lor abx) \lor (b\bar{x} \lor ab\bar{x}) &= \quad \text{(7.9)} \\
ax \lor abx \lor b\bar{x} \lor ab\bar{x} &= \quad \text{(7.20)} \\
ax \lor b\bar{x} \lor abx \lor ab\bar{x} &= \quad \text{(7.8)} \\
ax \lor b\bar{x} \lor ab &= \quad \text{(7.18)}
\end{aligned}$$

Q.E.D.

The immediate consequence of this is that the consensus of two terms of an SOP expression may be ORed with the expression (thereby adding another term) without changing the value of the expression as a whole.

As an example, consider the following expression in SOP (but not minterm) form:

$$ab \lor \bar{b}\bar{c} \lor a\bar{b}c.$$

The truth table corresponding to this function is shown in Table 8.1. Note that the first two terms, ab and $\bar{b}\bar{c}$ have exactly one opposition, b. Thus their consensus is $a\bar{c}$. ORing this with the original expression gives us

TABLE 8.1

a	b	c	ab	$\bar{b}\bar{c}$	$a\bar{b}c$	$ab \lor \bar{b}\bar{c} \lor a\bar{b}c$
0	0	0	0	1	0	1
0	0	1	0	0	0	0
0	1	0	0	0	0	0
0	1	1	0	0	0	0
1	0	0	0	1	0	1
1	0	1	0	0	1	1
1	1	0	1	0	0	1
1	1	1	1	0	0	1

TABLE 8.2

a	b	c	ab	$\bar{b}\bar{c}$	$a\bar{b}c$	$a\bar{c}$	$ab \vee \bar{b}\bar{c} \vee a\bar{b}c \vee a\bar{c}$
0	0	0	0	1	0	0	1
0	0	1	0	0	0	0	0
0	1	0	0	0	0	0	0
0	1	1	0	0	0	0	0
1	0	0	0	1	0	1	1
1	0	1	0	0	1	0	1
1	1	0	1	0	0	1	1
1	1	1	1	0	0	0	1

$ab \vee \bar{b}\bar{c} \vee a\bar{b}c \vee a\bar{c}$, which is the same as the function we had before is shown in Table 8.2.

EXERCISE 8.1

A. Find the consensus of the following terms if one exists.

1. $a\bar{b}\,\bar{d} \vee c\bar{d}$.
2. $aef \vee bc\bar{e}$.
3. $w\bar{x}z \vee xy\bar{z}$.
4. $\bar{a}\,\bar{b}\,\bar{f} \vee acde$.
5. Draw Venn diagrams illustrating the consensus rule. That is, draw the Venn diagrams of $a \cap x$, $b \cap x'$, and $a \cap b$ and show that $(a \cap x) \cup (b \cap x') = (a \cap x) \cup (b \cap x') \cup (a \cap b)$.
6. Give the dual of the consensus rule.
7. Put the following expression in SOP form, find all terms that have consensuses, and OR the consensuses onto the expression. Verify with a truth table that the expression you end with is equivalent to the one you started with.

$$[bd \wedge (a \vee \bar{c})] \vee \bar{a} \vee a\bar{d}$$

8.3 CANONICAL FORM

There are many different but equivalent realizations of the same function, including the minterm and maxterm realizations. Even if we restrict ourselves to SOP expressions, often we may simplify a minterm expression algebraically to the extent that while we still have an SOP expression, it is not immediately obvious that it represents the same function as the original minterm expression.

While there is no "right" realization of any given function, simpler realizations are certainly easier to understand and work with than complex realizations. It is convenient, moreover, to have one standard way of realizing any given Boolean function. If we always write Boolean expressions in a certain form, it will always be obvious, given two expressions, whether we are dealing with the same function or different functions. Such a form is called a **canonical form.** A political, academic, or religious canon is a body of literature that is considered official doctrine, which can be referred to for guidance. Similarly, a canonical form is an "official" form for writing algebraic expressions of a given type (Boolean algebraic expressions in our case).

For the purposes of determining whether a Boolean expression is in canonical form, we will not worry about differences between expressions that could be resolved by an application of either commutative rule. That is, we shall consider $ab\bar{c} \vee \bar{b}c$ to be the same expression as $c\bar{b} \vee \bar{c}ba$ (we shall tend, however, to keep the variables within a term in alphabetical order).

How do we choose a particular way of writing expressions to be our canonical form and invent a method for converting noncanonical expressions that form? First, we must be absolutely certain that each function has a unique representation in the canonical form we choose. That is, there is to be one and only one canonical realization for each function. It should be impossible to have two (or more) different canonical realizations of the same function (except differences due to the commutativity caveat mentioned above).

Why not just use minterm form? We already know that there is only one minterm realization of any function. This is a perfectly good candidate for canonical form, and some Boolean mathematicians use minterm form as their canonical form and have no problem with it. However, we have seen that minterm expressions often can be more complicated than one would think they should be, especially when the function they represent has more 1's than 0's in its truth table (remember the unwieldy minterm realization of the simple two-input OR function, $\bar{x}y \vee x\bar{y} \vee xy$). People who use minterm form as their canonical form often work around this awkwardness by using the shorthand notation for minterms (e.g., $M_1 \vee M_2 \vee M_3$). Nonetheless we can find a leaner canonical form.

Why not put all functions in SOP form of one kind or another and simplify until no further simplification is possible? It is reasonable to expect that no matter how different two functions look at first, once we whittle them down to their bare essentials algebraically, their differences or similarities ought to become apparent. However, this is not quite true. Consider the function $\bar{b}\bar{c} \vee bc \vee \bar{a}b$ whose output is shown in the truth table shown in Table 8.3.

Note that this same function is realized by the equally simple $\bar{b}\bar{c} \vee bc \vee \bar{a}\bar{c}$. Neither of these different expressions can be simplified any further; yet they represent the same function. Thus, a strategy of straight simplification does not

TABLE 8.3

a	b	c	$\bar{b}\bar{c} \vee bc \vee \bar{a}b$
0	0	0	1
0	0	1	0
0	1	0	1
0	1	1	1
1	0	0	1
1	0	1	0
1	1	0	0
1	1	1	1

yield a satisfactory canonical form for Boolean expressions because it cannot guarantee a unique realization of any given function.

8.4 BLAKE CANONICAL FORM

Blake showed in his 1937 dissertation ''Canonical Expressions in Boolean Algebra'' that a finite sequence of repetitions of the following two steps would produce a unique and compact realization of any given Boolean function that starts out in SOP form:

1. Simplification by condensing two terms into one by either of the absorption laws given in equations 7.14 and 7.16.
2. For any two terms with one opposition, ORing the consensus of the two terms onto the expression.

Once this sequence of steps has been performed on a given SOP expression until neither can be applied, the expression is said to be in **Blake canonical form** (BCF).

As an example, apply this method to find the Blake canonical form of the function whose truth table is given in Table 8.3. Start from scratch, with its minterm realization:

$$\bar{a}\bar{b}\bar{c} \vee \bar{a}b\bar{c} \vee \bar{a}bc \vee a\bar{b}\bar{c} \vee abc.$$

First, we ought to test each pair of terms to see if we can replace it with a single term according to the absorption laws. For simplicity, we will not mention applications of the commutative rule needed to put terms in exactly the form specified in the absorption rules. One way of accomplishing the term-by-term absorbability test systematically is by comparing each term in the expression to all the terms to its left. We begin with the second term ($\bar{a}b\bar{c}$), as the first has no terms to its left. After comparing the second term to the first (and finding no way to absorb them together), we compare the third term ($\bar{a}bc$) to the first and second terms, then the fourth term to the first, second, and third terms, and so on until every term in the expression has been compared to every other term. Unfortunately, we find that no pair of terms is absorbable on this first pass.

We then move on to step two, in which we again compare each pair of terms in the expression, looking for terms with exactly one opposition. If we find such a pair of terms, we OR their consensus onto the end of the expression. From then on, we treat this new term as we do all the other terms in the expression, including it in our term-by-term comparisons. This is why it is a good idea to start with each term in turn and compare it to all the terms to its left instead of the more obvious method of comparing each term to the terms to its right. This way, when we OR a new term (formed by the consensus of two already present terms) to the end of the expression during a pass, we are sure to end up testing it for absorbability with every other term in the whole expression before we are through.

Beginning by comparing the second term ($\bar{a}b\bar{c}$) to the first ($\bar{a}\bar{b}\bar{c}$), we see that they do have one opposition, b. Before we continue with the comparisons, we OR their consensus, ($\bar{a}\bar{c}$), onto the end of the expression:

$$\bar{a}\bar{b}\bar{c} \vee \bar{a}b\bar{c} \vee \bar{a}bc \vee a\bar{b}\bar{c} \vee abc \vee \bar{a}\ \bar{c}.$$

Hint: it is not cheating to absorb terms as you go along looking for terms that have a consensus. The outcome will be the same, and cutting down on the number of terms you have to deal with as you go along saves time. This said, it should be clear that the term we just added ($\bar{a}\ \bar{c}$) absorbs both of the terms whose consensus it is, namely the first two terms of the expression (according to equation 7.14). We may delete these terms:

$$\cancel{\bar{a}\bar{b}\bar{c}} \vee \cancel{\bar{a}b\bar{c}} \vee \bar{a}bc \vee a\bar{b}\bar{c} \vee abc \vee \bar{a}\bar{c}.$$

We continue to find pairs of terms that have one opposition. The third remaining term (abc) has the variable a in opposition to the first term ($\bar{a}bc$). Again, we may OR their consensus to the end of the expression and delete the (now absorbed) terms of which it is the consensus:

$$\cancel{\bar{a}\bar{b}\bar{c}} \vee \cancel{\bar{a}b\bar{c}} \vee \cancel{\bar{a}bc} \vee a\bar{b}\bar{c} \vee \cancel{abc} \vee \bar{a}\bar{c} \vee bc.$$

The second ($\bar{a}\bar{c}$) and first term ($a\bar{b}\bar{c}$) remaining have the variable a in opposition. ORing their consensus ($\bar{b}\bar{c}$) to the end gives us

$$\cancel{\bar{a}\bar{b}\bar{c}} \vee \cancel{\bar{a}b\bar{c}} \vee \cancel{\bar{a}bc} \vee a\bar{b}\bar{c} \vee \cancel{abc} \vee \bar{a}\bar{c} \vee bc \vee \bar{b}\bar{c}.$$

Note that this new term only absorbs one of the terms of which it is a consensus, $a\bar{b}\bar{c}$, but not $\bar{a}\bar{c}$. Deleting the absorbed term gives us

$$\bar{a}\cancel{\bar{b}}\bar{c} \vee \bar{a}\cancel{b}\bar{c} \vee \bar{a}\cancel{b}c \vee a\cancel{\bar{b}}\bar{c} \vee a\cancel{b}c \vee \bar{a}\bar{c} \vee bc \vee \bar{b}\bar{c}.$$

The only new term that can be added now by consensus is the consensus of the first ($\bar{a}\ \bar{c}$) and second (bc) remaining terms, giving us

$$\bar{a}\cancel{\bar{b}}\bar{c} \vee \bar{a}\cancel{b}\bar{c} \vee \bar{a}\cancel{b}c \vee a\cancel{\bar{b}}\bar{c} \vee a\cancel{b}c \vee \bar{a}\bar{c} \vee bc \vee \bar{b}\bar{c} \vee \bar{a}b.$$

A quick check verifies that no more terms can be absorbed, and no two terms have one and only one opposition. We are done, and our final expression is in Blake canonical form:

$$\bar{a}\bar{c} \vee bc \vee \bar{b}\bar{c} \vee \bar{a}b.$$

This expression still gives us the output shown in Table 8.3; putting an expression in Blake canonical form in no way changes its value.

This is the same function that was our counterexample earlier that proved that simplification alone did not yield a satisfactorily unique canonical form. We showed that the function could be realized by the equally simple but different expressions:

$$\bar{b}\bar{c} \vee bc \vee \bar{a}b \text{ and } \bar{b}\bar{c} \vee bc \vee \bar{a}\bar{c}.$$

Note that the Blake canonical form of the function consists of all terms in both of the above simplified expressions ORed together (with duplicate terms omitted). Furthermore, either of the above expressions can be easily put in Blake canonical form by applying Blake's two-step method. In the next section we will explore some of the peculiar properties of the Blake canonical form and make use of its applicability to logic problems.

EXERCISE 8.2

A. Put the following expressions in BCF (for some of them you must first put them in SOP form).

1. The minterm realization of the NAND function: $\bar{x}\bar{y} \vee \bar{x}y \vee x\bar{y}$.
2. $a\bar{c}\bar{d} \vee bc \vee \bar{a}\bar{b}d \vee \bar{c}d$.
3. $\bar{x}\bar{y}\bar{z} \vee \bar{x}\bar{y}z \vee x\bar{y}\bar{z} \vee x\bar{y}z$.
4. $\overline{(\bar{a} \vee b \vee \bar{c})} \vee [c \wedge (\bar{a}d \vee ab)]$.
5. The sum function from one cell of a cascading adder in terms of a, b, and c_i *(carry in):* $\bar{a}\bar{b}c_i \vee \bar{a}b\bar{c}_i \vee a\bar{b}\bar{c}_i \vee abc_i$.
6. The carry out function from one cell of a cascading adder in terms of a, b, and c_i (carry in): $\bar{a}bc_i \vee a\bar{b}c_i \vee ab\bar{c}_i \vee abc_i$.

8.5 PRIME IMPLICANTS

Because any Boolean expression implies itself ORed with any other Boolean expression (that is, $x \rightarrow (x \vee y)$ is always true), we know that any term of an SOP expression of a Boolean function implies or is an implicant of that function.

A term that is a **prime implicant** of a function is a ''minimal'' implicant: it has the property that if it were made any simpler by the deletion of any of its variables, it would no longer be an implicant of the function. Consider the function specified by the truth Table 8.4.

TABLE 8.4

x	y	z	
0	0	0	1
0	0	1	1
0	1	0	0
0	1	1	1
1	0	0	0
1	0	1	1
1	1	0	0
1	1	1	1

This function is realized by the SOP expression $x\bar{y}z \vee yz \vee \bar{x}\bar{y}$. The terms $x\bar{y}z$, yz, and $\bar{x}\bar{y}$ are all implicants of this function: if any of them is 1, the function will be 1. However, only $\bar{x}\,\bar{y}$ is a prime implicant. The function is 1 whenever $\bar{x}\bar{y}$ is 1 (i.e., whenever $(x, y) = (0, 0)$) but not whenever $\bar{x}$ alone is 1 or whenever $\bar{y}$ alone is 1. Thus, the term $\bar{x}\,\bar{y}$ is as compact as it can be while still being an implicant of this function. If we look closely at the truth table, however, we see that the function is 1 whenever z is 1. Therefore z itself is an implicant of this function. The two terms that contain the variable z ($x\bar{y}z$ and yz) are thus not prime implicants because each of them could be stripped of all variables except z (in which case they would be identical, consisting solely of the variable z) and they would still be implicants of the function.

An expression in Blake canonical form has the useful property of consisting entirely of prime implicants ORed together. In fact, a function expressed in Blake canonical form is guaranteed to contain *all* of the prime implicants of that function. We saw that $\bar{b}\bar{c} \vee bc \vee \bar{a}b$ and $\bar{b}\bar{c} \vee bc \vee \bar{a}\bar{c}$ both realize the same Boolean function. Both consist of prime implicants ORed together. However, each contains a prime implicant that the other does not. The Blake canonical form of the function, on the other hand, contains all prime implicants, $\bar{a}\bar{c} \vee bc \vee \bar{b}\bar{c} \vee \bar{a}b$.

The process of putting an SOP expression into Blake canonical form is for the most part one of simplication. We generally begin with lengthy expressions whose terms contain many variables and we condense them into a few smaller terms. Through absorption we delete redundant variables and are left with an expression consisting solely of prime implicants, which tell us as simply as possible exactly what it takes to make the function 1. It is worth noting, however, that in the case just described, the Blake canonical form of the function is not the simplest realization. Both of the realizations we started with were shorter. The Blake canonical form of a function is generally a compact realization, but it is guaranteed to contain all of the prime implicants of a function, even those that are redundant with combinations of some of the other prime implicants.

As appealing as the promise of simplicity is, to logicians it is this second aspect of the Blake canonical form that is at least as useful as its simplifying properties (if not more so). Specifically, a systematic method for finding all prime implicants of a function will enable us to ferret out hidden implicants and, therefore, hidden logical conclusions from a given set of premises. The power of this method, when applied to the field of logic, is something even Aristotle himself could not have imagined.

8.6 A NEW REALIZATION OF THE IMPLICATION FUNCTION

There is one more brief detour we must take before we jump into the heart of this chapter. This detour involves deriving a different realization of the implication function than the one we have been using, namely $x \rightarrow y = \bar{x} \vee y$. This new realization will be useful to us shortly. To say that x implies y is to say that whenever x is 1, y is also 1. Put another way, there is no way x could be 1 while y is 0. In Boolean terms, this verbal understanding of the implication function becomes the expression $x \wedge \bar{y} = 0$: x AND NOT y equals the constant 0 function. The principle of assertion tells us that an expression set equal to 0 is equivalent to its own complement, so $x \wedge \bar{y} = 0$ may be formulated as the expression $\overline{(x \wedge \bar{y})}$. Using DeMorgan's laws we can evaluate this using the following equations:

$$\overline{(x \wedge \bar{y})} =$$

$$\bar{x} \vee \bar{\bar{y}} = \qquad (7.23)$$

$$\bar{x} \vee y. \qquad (7.7)$$

As $\overline{(x \wedge \bar{y})}$ is equal to the more familiar realization of the implication function in terms of AND, OR, and NOT, namely $\bar{x} \vee y$, we have another realization of the implication function:

$$x \rightarrow y = \overline{x \vee \bar{y}}. \qquad (8.2)$$

8.7 SYLLOGISTIC REASONING

The term **syllogism** refers to a form of reasoning that is fundamental to Aristotelian logic and that is best exemplified by the classic argument, "All men are mortal. Socrates is a man. Therefore Socrates is mortal." The conclusion ("Socrates is mortal'') is derived from the two premises (''All men are mortal'' and ''Socrates is a man''). While this is a simple example, deriving such logical conclusions from a set of premises can be fiendishly complex. In this section we will learn how to frame word problems in symbolic terms that we can manipulate according to the rules of Boolean algebra, allowing us to extract whatever conclusions are logically deducible from the premises. The method by which we will accomplish this is known as **syllogistic reasoning.**

To use syllogistic reasoning, we begin by phrasing the premises in terms of Boolean algebra with each premise forming a separate Boolean expression. For instance, let us translate the premises of the above syllogism into a pair of implications, selecting variables as follows: m = mortal, h (for human, so as not to conflict with the m for mortal) = man, s = Socrates.

$$\text{All men are mortal:} \qquad h \rightarrow m.$$

$$\text{Socrates is a man:} \qquad s \rightarrow h.$$

The use of the implication function follows from its set algebraic interpretation ($x \subset y$) and slightly rephrasing the premises as ''The set of men is a subset of the set of mortals,'' and ''The set of Socrates is a subset of the set of men.'' Alternately, we may use the Boolean logic interpretation of the implication and rephrase the premises as ''X is a man implies that X is mortal'' and ''X is Socrates implies that X is a man.'' Either way, it should be clear that the essence of the two premises of the syllogism is carried in the Boolean expressions $h \rightarrow m$ and $s \rightarrow h$.

Having put each of the premises in the form of a Boolean expression, we now convert these into Boolean expressions that are equal to 0. We do this by applying equation 8.2, the alternative realization of the implication function in terms of AND, OR, and NOT, which followed so neatly from the logical interpretation of the implication function.

According to the principle of assertion, the expression $\overline{(x \wedge \bar{y})}$ is the same as the statement $(x \wedge \bar{y}) = 0$ (i.e., there is no way x could be 1 and y could be 0). Throughout this book we have avoided such equations, preferring to say ''$x \wedge \bar{y}$'' instead of ''$x \wedge \bar{y} = 1$'' and ''$\bar{x}$'' instead of ''$x = 0$.'' Boole worked a lot with equations, however, and for the purposes of syllogistic reasoning, they provide a clearer sense of the steps involved.

Here are our premises, now translated into expressions set equal to 0:

$$h\bar{m} = 0.$$

$$s\bar{h} = 0.$$

Now we OR all our premises together into one big expression. Since the premises are all equal to 0, this expression will be equal to 0 as well, since $0 \vee 0 = 0$:

$$h\bar{m} \bigvee s\bar{h} = 0.$$

This gives us an SOP expression that is equal to 0 and that we now convert to Blake canonical form. The two terms in this equation have an opposition, h. We OR their consensus, $s\bar{m}$, to the end of expression that, as we know, will not change the value of the expression as a whole:

$$h\bar{m} \bigvee s\bar{h} \bigvee s\bar{m} = 0.$$

No terms can be absorbed and no pair of terms has an opposition, so we are done. Now we simply reverse the process that got us this far. Our knowledge of the OR function tells us that if an SOP expression is equal to 0, each of its terms must also be equal to 0 (if any of them were equal to 1, the whole expression would equal 1). Setting the individual terms equal to 0, translating back into statements involving the implication function by way of equation 8.2, and then to English sentences yields the following:

$h\bar{m} = 0$	$h \rightarrow m$	"All men are mortal."
$s\bar{h} = 0$	$s \rightarrow h$	"Socrates is a man."
$s\bar{m} = 0$	$s \rightarrow m$	"Socrates is mortal."

The first two are our original premises, unchanged. The third was revealed by converting the expression formed by the first two into Blake canonical form. Blake canonical form is guaranteed to reveal all conclusions that may be logically inferred from a given set of premises, even when those conclusions are not at all immediately obvious from a cursory reading of the premises. This is the magic of syllogistic reasoning. The only real trick to this method is translating premises from English sentences to Boolean expressions equal to 0 and then translating terms of the final expression back into English.

Boole worked quite a bit with converting systems of Boolean equations to one big expression set equal to 0 to arrive at logical conclusions, but Blake developed the algorithm that we now use to find prime implicants. The term *prime implicant,* however, was coined by Willard Van Orman Quine in 1952. Whereas Blake approached Boolean algebra from the time-honored perspective of logic, Quine independently developed similar techniques but applied them to the different problem of formula minimization with an eye toward circuit simplification (remember, this was the 1950s when individual logic gates were quite pricey).

8.8 PREMISES THAT ARE NOT IMPLICATIONS

We saw in Section 8.7 how easily a premise that takes the form of an implication can be converted to an SOP Boolean expression that equals 0. Premises can take other forms, however. They can be equalities (e.g., "All men are mortals and vice versa'', $h = m$). In such a case, to convert the premise to a Boolean equation equal to 0, we use the following translational rule:

$$h = m \text{ becomes } h\bar{m} \vee \bar{h}m = 0.$$

This can be justified by the fact that if $h = m$, then $h \oplus m = 0$, and $h \oplus m$ in SOP form is $h\bar{m} \vee \bar{h}m$. This has the logical consequence that ''$x = y$'' and ''$x \rightarrow y$ and $y \rightarrow x$'' are equivalent statements.

This translation yields not one term equal to 0, but two terms ORed together. This is fine, as the two terms still must individually be equal to 0. The method works as long as all premises can be ORed together into an SOP expression that is equal to 0 regardless of how many terms the individual premises have.

A premise may also take the form of an inequality or mutual exclusion (e.g., ''No men are mortal, and all nonmen are mortal'', $h \neq m$). In this case, we use the fact that if $h \neq m$, then $(h = m) = 0$. This leads to the following translational rule:

$$h \neq m \text{ becomes } \bar{h}\bar{m} \vee hm = 0.$$

This is justified by the fact that the equality function (coincidence) is defined by the SOP expression $\bar{h}\bar{m} \vee hm$.

To summarize, applying syllogistic reasoning involves the following steps:

1. Express each premise as a term or SOP expression set equal to 0. Premises that take the form of implications may be so translated according to the rule $x \rightarrow y = \overline{(x\bar{y})}$, or $x\bar{y} = 0$ according to the principle of assertion.
2. Link all resulting terms and small SOP expressions together with ORs, thus forming one large SOP expression, which is set equal to 0.
3. Put this SOP expression in Blake canonical form.
4. Set each term from the BCF expression (i.e., the prime implicants) equal to 0 separately.
5. Interpret each resulting equation in the form of an English sentence.

We will now try a slightly more complicated example. Imagine that you are watching sports on television with a small group of friends, and you have been selected to take food orders and make a quick trip to the local convenience store. Immediately, your friend Alex pipes up, ''This time, remember to get salsa if you get nachos, OK?'' to which Denise adds, ''All you guys ever eat is nachos and salsa. If ice cream sandwiches are on sale, get them and skip the nachos.'' Confident that you have this straight in your mind, you head for the door. But just as you are putting on your coat, your host pulls you aside and says, ''If you get some ice cream sandwiches and some milk, get some salsa as well just in case. Otherwise forget about the salsa.'' It takes you halfway to your car to stop wondering what your host wants to do with ice cream sandwiches, milk, and salsa and realize that you are completely confused.

Your orders concerned four different foods, so let us choose four appropriate variables to represent each: n for nachos, s for salsa, i for ice cream sandwiches, and m for milk. These variables are bits, and they represent your purchase or nonpurchase of the corresponding foods (and I use the term loosely). Thus, the variable i means you buy ice cream sandwiches, $\bar{i}$ means you do not. The instructions, then, can be represented as follows:

$n \rightarrow s$	If you get nachos, get salsa.
$i \rightarrow \bar{n}$	If you get ice cream sandwiches, don't get nachos.
$im = s$	Get salsa if and only if you get ice cream sandwiches and milk.

The first two expressions are straightforward implications and can be turned into expressions equal to 0 according to the rule: $x \rightarrow y = \overline{(x\bar{y})}$. The third is an equality, however, and must be broken down into exclusive OR as follows:

$$n\bar{s} = 0$$

$$in = 0$$

$$im \oplus s =$$

$$im\bar{s} \vee \overline{(im)}\, s =$$

$$im\bar{s} \vee [(\bar{i} \vee \bar{m}) \wedge s] =$$

$$im\bar{s} \vee \bar{i}s \vee \bar{m}s = 0.$$

We had to apply DeMorgan's laws and the distributive law to put the XOR expression, $im\bar{s} \vee \overline{(im)}s$, in SOP form. Note that we are implicitly applying the commutative rule when necessary to keep the variables in alphabetical order within each term. This will make it easier to compare terms to find oppositions and absorbable terms.

ORing all of the expressions together yields

$$n\bar{s} \vee in \vee im\bar{s} \vee \bar{i}s \vee \bar{m}s = 0.$$

Reducing to Blake canonical form gives us

$$im\bar{s} \vee \bar{i}s \vee \bar{m}s \vee n = 0.$$

Broken into separate expressions, all equal to 0, yields

$$im\bar{s} = 0$$

$$\bar{i}s = 0$$

$$\bar{m}s = 0$$

$$n = 0.$$

We know that the first three equations are derived from the condition $im = s$, the Boolean equivalent of the instruction to buy salsa if and only if you also got milk and ice cream sandwiches. However, they may be interpreted separately as implications as follows:

$$im\bar{s} = (im)\bar{s} = 0 \quad \Rightarrow \quad im \rightarrow s$$

$$\bar{i}s = s\bar{i} \quad \Rightarrow \quad s \rightarrow i$$

$$\bar{m}s = s\bar{m} \quad \Rightarrow \quad s \rightarrow m.$$

These effectively say the same thing: if you buy milk and ice cream sandwiches, also buy salsa, and if you buy salsa, buy milk, and if you buy salsa, buy ice cream sandwiches. The fourth equation is of particular interest. It says simply, $n = 0$. Don't buy nachos! While it is not immediately obvious from the instructions in the form in which they were given, it is now clear that there is no way you could buy nachos and satisfy your friends' wishes concerning your shopping expedition.

8.9 CLAUSAL FORM

At this point a word should be said about step five in our procedures of syllogistic reasoning, namely interpreting the results of a logical expression once it is in Blake canonical form. At least some of the original premises often are already prime implicants, and thus are bound to show up in the final expression in Blake canonical form. If one of the original premises is not itself a prime implicant, then it is absorbed by a term that is a prime implicant, and thus is present in the final expression.

Once we have an expression in Blake canonical form consisting of terms ORed together, we set each term equal to 0 separately. Sometimes these terms contain only two variables, one of which is negated, in which case the term can be easily turned into an implication ($x\bar{y} = 0$ tells us that $x \rightarrow y$). Sometimes, as in the previous example, one of the terms in the final BCF expression will be only a single variable. In that case, interpretation is also easy (e.g., $\bar{x} = 0$ tells us that $x = 1$).

Much of the time, however, logical interpretation of the separate prime implicants set equal to 0 will not be quite so straightforward. Often they will be a jumble of variables, some complemented, some not complemented. All such terms can be put in the form $x\bar{y}$, however, and thus interpreted as an implication. A term that has been manipulated in this way is said to be in **clausal form.** We will categorize the possible forms a term can take into three cases and see how each of them can be interpreted.

Case 1. *The term consists of several variables, at least one complemented and at least one not complemented.*

That is, the term is a mixture of complemented and uncomplemented variables. We use the commutative rule to move all the uncomplemented variables to the left and all the complemented variables to the right. Then we group the complemented variables together with parentheses and use the DeMorgan Law to turn this grouping from a collection of complemented variables ANDed together to the complement of a collection of uncomplemented variables ORed together. The entire term now can be thought of as a single implication in which the AND of the uncomplemented variables is the implicant and the OR of the rest of the variables is the consequent.

For example, consider a case in which one of the prime implicants in an expression in Blake canonical form is $\bar{b}c\bar{d}\bar{e}f$. This gives us the equation $\bar{b}c\bar{d}\bar{e}f = 0$. To interpret this, we commutatively shuffle the variables so the complemented ones are all on the right and the uncomplemented ones are all on the left:

$$cf\bar{b}\bar{d}\bar{e} = 0.$$

We group the complemented variables together as allowed by the associative law of AND:

$$cf \wedge (\bar{b}\bar{d}\bar{e}) = 0$$

and apply DeMorgan's law:

$$cf \wedge \overline{(b \vee d \vee e)} = 0.$$

The equation can be interpreted as $cf \rightarrow (b \vee d \vee e)$ or "If c and f, then b or d or e."

Case 2. *All variables in the term are uncomplemented.*

The term ab, for example, gives us the equation $ab = 0$. Using the involution rule, we could write this as $a(\bar{\bar{b}})$, which could be interpreted as $a \rightarrow \bar{b}$, but we could just as easily interpret the term as $b \rightarrow \bar{a}$. The clearest way of interpreting such a result is to realize that $ab = 0$ is the same as $\overline{(ab)}$, the NAND function. Thus, the English interpretation of $ab = 0$ is "It is not the case that both a and b." This is extendable to any number of uncomplemented variables ANDed together.

The clausal form of the term $ab = 0$ is derived by ANDing the term with 1 (which of course leaves its value unchanged):

$$ab = 0$$

$$ab \wedge 1 = 0$$

$$ab \wedge \bar{0} = 0.$$

This, finally, may be put in the form of an implication, namely $ab \rightarrow 0$. This implication has the interpretation, "whenever a and b are 1, then 0 is 1," which is never, so a and b cannot be 1 at the same time.

Case 3. *All variables in the term are complemented.*

In this case, we group the variables in the term together and apply DeMorgan's law. For example, suppose we are left with the prime implicant $\bar{w}\bar{y}\bar{z}$, giving us the equation $\bar{w}\bar{y}\bar{z} = 0$.

$$\bar{w}\bar{y}\bar{z} = 0$$

$$(\bar{w}\bar{y}\bar{z}) = 0$$

$$\overline{(w \vee y \vee z)} = 0$$

We now have a complemented expression set equal to 0. We may derive an equivalent expression by complementing each side of the equation to give us

$$(w \vee y \vee z) = 1$$

the English interpretation of which is simply "w, or y, or z."

The clausal form of this case is derived by ANDing the expression with 1 as follows:

$$\bar{w}\bar{y}\bar{z} = 0$$

$$\overline{(w \vee y \vee z)} = 0$$

$$1 \wedge \overline{(w \vee y \vee z)} = 0.$$

This is equivalent to the implication $1 \rightarrow (w \vee y \vee z)$, "whenever 1 is 1, then so is $(w \vee y \vee z)$."

8.10 VIDEO GAME EXAMPLE

As a further example, let us imagine that you are in a video arcade and are about to play a new game. Before you put your coin into the slot, however, you break tradition and actually read the little description of the game as it scrolls up the screen over an animated backdrop of graphic violence amid postapocalyptic urban rubble:

> It is the year 2105. A global war between two rival energy conglomerates has left the Earth a burnt-out shell. The world's leaders, lead by the international banking

cartels, have abandoned Earth for an off-world colony. Before they left, however, they installed their minions, consisting of robots and grotesque genetically engineered mutants, to rule in their place over what is left of society.

You, the leader of the remaining humans, must reclaim the planet from their tyrannical grip! As you make your way through the labyrinthine corridors of their central compound toward the hub of their communications network, remember:

Some minions guarding the hub are armored and impervious to your phased plasma gun. Some are capable of teleportation. Some are vulnerable to your helmet-mounted electro-neural disrupter field.

1. No minions can teleport while wearing armor.
2. Any minion worth 150 points but still vulnerable to the electro-neural disrupter can teleport.
3. All 150-point minions are armored.
4. Any minion that is electro-neurally disruptable and not worth 150 points is either armored or a teleporter or both.
5. Any minion that is invulnerable to both of your weapons either can teleport or is not worth 150 points or both.

* Good luck! The fate of humanity rests in your hands!*

0 credits

Insert coin

How may we simplify the game's elaborate scoring system? This example differs in a subtle way from the last one. While we will approach it using the same techniques of Boolean algebra, it is formulated in terms of the algebra of sets rather than strict Boolean implications. Whereas before we were concerned with such statements as "If you do this, then don't do that or that," here we are concerned with sorting out overlapping categories of evil minions. The problem may be conceptualized either way: rule three could be stated either as "the set of 150-point minions is a subset of the set of armored minions" or "if a minion is worth 150 points, then it is armored." For the sake of variety, however, we will approach this using the terminology of the algebra of sets, even though we shall use the familiar Boolean symbols when manipulating expressions.

We must first assign variables to the sets involved as follows:

w: set of minions worth 150 points.

a: set of armored minions.

t: set of minions that can teleport.

d: set of minions that are electro-neurally disruptable.

We must state each scoring rule algebraically as an expression set equal to 0. As before, many of them will first take the form of expressions involving the implication function, then be set equal to 0 according to the rule $x \rightarrow y = \overline{(x\bar{y})}$. Although we conceptualize 0 as being the null set and the implication function as meaning "is a subset of," the algebraic manipulations are similar to those used in the previous example.

Here are our five rules, reworded and stated algebraically as expressions set equal to 0, with intermediate steps shown:

1. The intersection of armored minions and teleporting minions is null:

$$a \wedge t = 0.$$

2. The intersection of the set of minions that are worth 150 points and the set of minions that are vulnerable to the electro-neural disrupter is a subset of the set of teleporting minions:

$$(w \wedge d) \rightarrow t$$

$$w \wedge d \wedge \bar{t} = 0.$$

3. The set of 150-point minions is a subset of the set of armored minions:

$$w \rightarrow a$$

$$w \wedge \bar{a} = 0.$$

4. The intersection of the set of electro-neurally disruptable minions and the complement of the set of 150-point minions is a subset of the union of the sets of armored and teleporting minions:

$$(d \wedge \bar{w}) \rightarrow (a \vee t)$$

$$(d \wedge \bar{w}) \wedge \overline{(a \vee t)} =$$

$$d \wedge \bar{w} \wedge \bar{a} \wedge \bar{t} = 0.$$

5. The intersection of the set of armored minions and the complement of the set of minions that is electro-neurally disruptable is a subset of the union of the teleporting minions and the complement of the set of 150-point minions:

$$(a \wedge \bar{d}) \rightarrow (t \vee \bar{w})$$

$$(a \wedge \bar{d}) \wedge \overline{(t \vee \bar{w})} =$$

$$a \wedge \bar{d} \wedge \bar{t} \wedge \bar{\bar{w}} =$$

$$a \wedge \bar{d} \wedge \bar{t} \wedge w = 0.$$

This yields the following set of equations:

$$a \wedge t = 0$$

$$w \wedge d \wedge \bar{t} = 0$$

$$w \wedge \bar{a} = 0$$

$$d \wedge \bar{w} \wedge \bar{a} \wedge \bar{t} = 0$$

$$a \wedge \bar{d} \wedge \bar{t} \wedge w = 0.$$

Thus, the single equation:

$$at \vee d\bar{t}w \vee \bar{a}w \vee \bar{a}d\bar{t}\bar{w} \vee a\bar{d}\bar{t}w = 0.$$

Reduced to Blake canonical form, this becomes

$$at \vee \bar{a}d\bar{t} \vee w = 0,$$

which is equivalent to the following three separate expressions:

$$a \wedge t = 0$$

$$\bar{a} \wedge d \wedge \bar{t} = 0$$

$$w = 0.$$

Note that the prime implicant $a \wedge t$ was present in the original expression. It could be interpreted verbally as it originally was presented by the video game (''No minions can teleport while wearing armor''). We will use the techniques of the last section, however, to interpret it in clausal form.

The equation $a \wedge t = 0$ is an example of case 2 (a term in which there are no complemented variables) that could be interpreted as, ''It is not the case that a minion can wear armor and teleport.'' Clausally, this may be interpreted as $a \wedge t \rightarrow 0$, or ''The intersection of the set of armored minions and teleporting minions is a subset of the null set,'' which is another way of saying that the intersection is empty.

The equation $\bar{a} \wedge d \wedge \bar{t} = 0$ is an example of case 1, in which the prime implicant consists of a mix of complemented and uncomplemented variables. Using DeMorgan's law it may be put in clausal form through the following manipulation:

$$\bar{a} \wedge d \wedge \bar{t} =$$

$$d \wedge \bar{a} \wedge \bar{t} =$$

$$d \wedge (\bar{a} \wedge \bar{t}) =$$

$$d \wedge \overline{(a \vee t)} = 0.$$

This, according to equation 8.2, is equivalent to the expression

$$d \rightarrow (a \vee t).$$

This may be interpreted as, ''The set of electro-neurally disruptable minions is a subset of the union of the sets of armored and teleporting minions.''

The final equation is of particular interest. It says that $w = 0$. This is an example of a case in which all variables in a term are uncomplemented, even though there is only one of them. Interpreting it is simple: the set of all minions worth 150 points is null. That is, there are no minions at all that are worth 150 points. Therefore, our original five rules may be condensed into three with no loss of information:

1. A minion cannot wear armor and teleport.
2. All electro-neurally disruptable minions are either armored or can teleport (or both).
3. No minions are worth 150 points.

8.11 EQUIVALENT FORMULATIONS OF RESULTS

While we have adopted the clausal form described previously as our method for converting a prime implicant to an implication expression and then to an English statement, there are equivalent variations of this form. Specifically, it is not necessary to group complemented and uncomplemented variables as previously described.

Consider the term $\bar{a}d\bar{t}$ from the last example. According to the rules of clausal form, this is equivalent to the statement $d \rightarrow (a \vee t)$. However, we could have interpreted the term as $(d \wedge \bar{t}) \wedge \bar{a} = 0$. This is equivalent to the implication $d\bar{t} \rightarrow a$ (All minions that are electro-neurally disruptable and cannot teleport are armored). This implication is equivalent to $d \rightarrow (a \vee t)$, which we derived above, as are the similarly derived $\bar{a}d \rightarrow t$, $\bar{a}\bar{t} \rightarrow \bar{d}$, and $\bar{t} \rightarrow (a \vee \bar{d})$.

Essentially, the prime implicant itself set equal to 0 is considered to be an implication decomposed into AND, OR, and NOT, which is then reconstituted into an implication statement in any of several different ways, all equivalent. The clausal form described previously however, tends to be the most straightforward representation of the information contained in the prime implicant as it contains the fewest possible complements.

8.12 SPECIFIC RESULTS DERIVED FROM GENERAL RESULTS

Given the three scoring rules we derived from the original five that the video game gave us, can we conclude that any armored minion either cannot teleport or is electro-neurally disruptable? Are all teleporting minions invulnerable to your disrupter field? To determine the answer to such specific questions, we phrase each as a proposition that we test by putting it in the form of an SOP expression in Blake canonical form set equal to 0. The proposition is true if the resulting expression can be shown to be equal to 0 given what we already know about the variables involved.

For example, the question, "Can we conclude that any armored minion either cannot teleport or is electro-neurally disruptable?" may be rephrased as the proposition that any armored minion either cannot teleport or is electro-neurally disruptable. This proposition can be represented as the implication $a \rightarrow (\bar{t} \vee d)$. This is equivalent to the equation $a \wedge \overline{(\bar{t} \vee d)} = 0$, which may be transformed to a term composed only of variables ANDed together by a sequence of the following familiar manipulations:

$$a \wedge \overline{(\bar{t} \vee d)} = 0$$

$$a \wedge (\bar{\bar{t}} \wedge \bar{d}) = 0$$

$$a \wedge (t \wedge \bar{d}) = 0$$

$$a\bar{d}t = 0.$$

The three rules that we derived about the video game are represented by the equation $at \vee \bar{a}d\bar{t} \vee w = 0$, so we must test the new assertion ($a\bar{d}t = 0$) against this expression. For the term being tested to be equal to 0, it must be absorbable into some term of the expression against which it is being tested. This is a consequence of the fact that this expression is in Blake canonical form and thus is composed of all its possible prime implicants. Prime implicants are the simplest, most general implicants of an Boolean function, and as such, all nonprime implicants of an expression imply some prime implicant of it.

As it happens, the term being tested, $a\bar{d}t$, is absorbed by the first term of the expression $at \vee \bar{a}d\bar{t} \vee \bar{w}$. That is, since we know that $at = 0$, then

$$a\bar{d}t =$$

$$at \wedge \bar{d} =$$

$$0 \wedge \bar{d} =$$

$$0.$$

Thus the assertion $a \rightarrow (\bar{t} \vee d)$ is true. However, the proposition that all teleporting minions are invulnerable to your disrupter field is represented by the implication $t \rightarrow \bar{d}$, which is equivalent to the expression $t\bar{\bar{d}} = td = 0$. The term td is not absorbable into any of the three terms of the expression $at \vee \bar{a}d\bar{t} \vee w = 0$, so it cannot be said with certainty that $td = 0$ given the facts that we have. Thus, some teleporting minions might be quite disruptable indeed.

If a term being tested is not readily absorbed into some prime implicant of the expression against which it is being tested, then it is not necessarily equal to the constant 0, and the implication that it represents does not necessarily hold true. It is not necessarily false either, but it is not a logical consequence of the facts already established.

EXERCISE 8.3

A. State in common English the conclusions that may be logically inferred from each of the following sets of premises.

1. Anyone who does not ride a hover scooter must not be a resident of Cathode City. Anyone who does not own a jet pack must ride a hover scooter. Anyone who has a jet pack and is eerily happy must live in Cathode City.
2. All murpels are also gleebs, but no gleeb is also a phaetino. All nonmurplels are phaetinos, however.
3. If the carpet is red, then the wallpaper must not be orange. If the sofa is striped, then the walls must be orange. If the carpet is not red, then the sofa must not be striped.
4. Whenever the buzzer sounds and the light flashes, a warning is displayed on the console. Whenever the light flashes, the bell rings. Whenever the warning is displayed, no bell is ringing.
5. All carnivores must be held in reinforced cages. All winged animals must live in well ventilated areas. If an animal must be bathed daily, it must not have wings. All herbivores have wings.

8.13 KARNAUGH MAPS

For functions of four of fewer variables, there is an easier, more mechanical way of finding prime implicants than putting the function in Blake canonical form through algebraic manipulation. This method involves visual inspection of the function's *Karnaugh map.*[1] A Karnaugh map of a function is just its truth table written slightly differently in the form of a grid or chart. Before we can draw a Karnaugh map, however, we must learn about gray code.

8.13.1 Gray Code

Gray code, or reflected code, is a way of writing out a sequence of binary numbers in such a way that only one bit ever ''flips'' from one number to the next. For example, when we count normally in base 2 (let us use three bits for the moment), we begin thus: 000, 001, 010, 011, 100. Only the last bit flipped

[1] This method was developed in full by M. Karnaugh in a paper he wrote in 1953, but his work was partly anticipated a year earlier by E. W. Veitch. For this reason, Karnaugh maps are sometimes named for Veitch or for both of them together.

from 000 to 001, but between 001 and 010, both the last bit and the middle bit flipped. Likewise, when 011 became 100, all three bits flipped. In gray code, counting all combinations of three bits would result instead in this sequence: 000, 001, 011, 010, 110, 111, 101, 100. Note that this is a significant departure from the way of thinking about numbers (and numerals) used in the Hindu-Arabic system. In gray code there is no concept of a distinct digit position that carries a certain weight that gets multiplied by the value of the numeral occupying that position.

In general, to write a sequence of gray code binary numbers consisting of n bits, you begin with n 0's. To generate the next number in the sequence, start with the previous number and flip the least significant bit you can without generating a number you have already generated somewhere else in the sequence. So, if we begin with 000, we immediately generated 001 as the second number in the sequence (we flipped the least significant bit). However, if we flip the least significant bit of 001, we will just get 000 back, which is already in the list. So we flip the middle bit and get 011. Now we may safely flip the least significant bit of this third entry in our list to get 010. To generate the fifth number, if we flip the least significant bit of 010, we just get back to 011, which is already in the list. Similarly if we flip the middle bit, we get 000, the very first entry. So we must flip the most significant bit to derive 110. Now we may flip the least significant bit of this to get our sixth entry, 111. We flip the middle bit to get the seventh entry, 101, and the least significant bit of this for our eighth and final number, 100.

Gray code is sometimes called **reflected code** because if you write a sequence of n-bit gray coded numbers in a vertical list and then strip off the most significant bit, the bottom half of the list will be a sequence of (n-1)-bit numbers that exactly reflects the sequence in the top half. This property is illustrated in Figure 8.1.

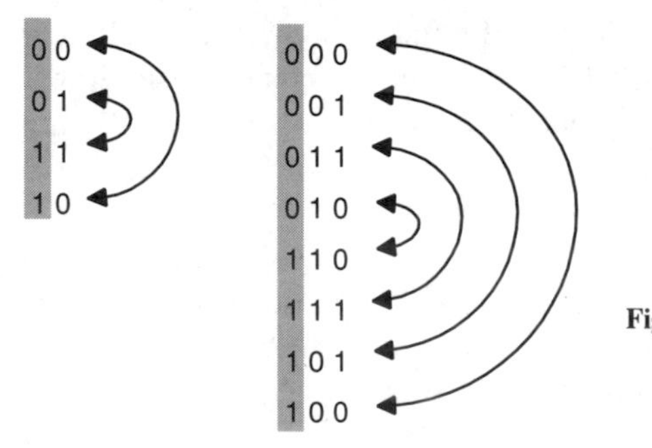

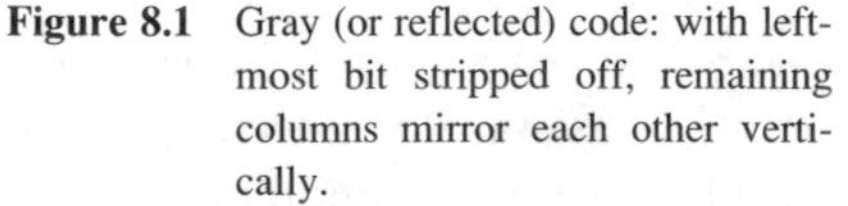
Figure 8.1 Gray (or reflected) code: with leftmost bit stripped off, remaining columns mirror each other vertically.

There are digital circuits that for one reason or another actually perform mathematical functions in gray code. These reasons usually have to do with the imperfect interface between the digital circuits involved and the real physical world. For example, if a digital sensor only flips one bit at a time as it registers consecutive values, there is never any danger of indeterminacies as the bits flip (one bit flipping an instant before the other or a bit in a high bit position fluttering between 0 and 1). For our purposes, however, it is enough to be able to write a sequence of two bit numbers in gray code.

To write a traditional truth table for a function of n bits, we write out all 2^n combinations of n bits (in strict ascending order) in a vertical list. Then we write either a 0 or a 1 to the right of each of the 2^n combinations, depending on whether that particular combination of the n input bits will make the function 0 or 1.

To generate a Karnaugh map, we write the table as a grid, with all possible combinations of some of the n input bits along the top and all possible combinations of the rest of the input bits along the side. These combinations of bits along the top and side are listed in gray code order. Thus, for a four input bit function, we would write all four combinations of the first two bits along the side in gray code order (creating four rows in the map) and all four combinations of the other two bits along the top in gray code order (creating four columns). For a three-input bit function, we have a choice of a four-row, two-column map or a two-row, four-column map.

Each of the resulting 2^n squares in the grid corresponds to a particular combination of the input bits of the function being mapped, just as each of the 2^n rows of the truth table did. The boxes in the grid are filled in with 1's as appropriate. Karnaugh maps (in two dimensions, anyway!) simply do not work if there are more than two of the input bits represented along either the top or the side, which is why they are only useful for functions of four bits or fewer.

As an example, let us revisit the function specified by the truth table shown in Table 8.3. We have seen that its minterm realization is

$$\bar{a}\bar{b}\bar{c} \bigvee \bar{a}b\bar{c} \bigvee \bar{a}bc \bigvee a\bar{b}\bar{c} \bigvee abc$$

and its realization in Blake canonical form is

$$\bar{a}\bar{c} \bigvee bc \bigvee \bar{b}\bar{c} \bigvee \bar{a}b.$$

Recall that the BCF realization of a function consists of all of that function's prime implicants ORed together. We now will find these same prime implicants using a Karnaugh map. As there are three input bits, a, b, and c, we can draw a grid with one bit along the side and two along the top or two bits along the side and one along the top. Let us arbitrarily pick the latter, putting a and b on the side and c on the top. We list all possible combinations of a and b in gray code order and write these combinations along the side, one combination per row, and all possible combinations of c alone along the top as shown in Figure 8.2.

Now we put 1's in those squares in the map that correspond to combinations of a, b, and c that result in the function being 1 according to the truth table shown in Figure 8.3.

Now, look at the map. Because we used gray code to list our combinations

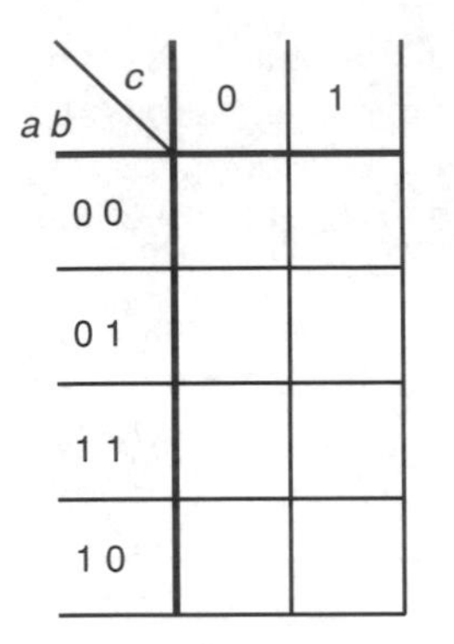

Figure 8.2 Blank Karnaugh map for a three-input function.

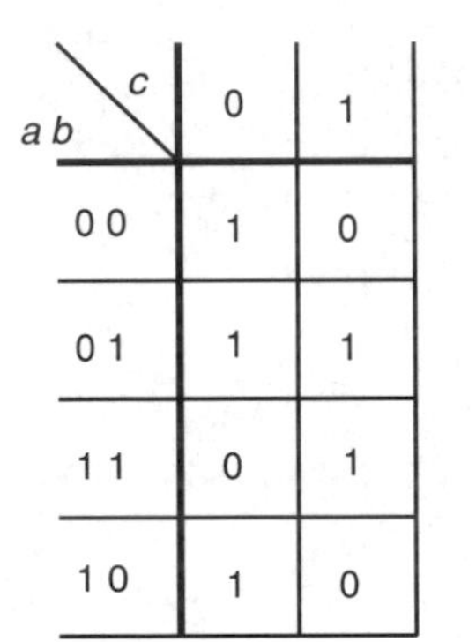

Figure 8.3 Karnaugh map for a three-input function with output bits filled in.

of input bits, any two vertically or horizontally adjacent squares differ only by one input bit. For our purposes, we may think of the table as wrapping around; that is, the top row is to be considered adjacent to the bottom row, and the rightmost column is considered to be adjacent to the leftmost column. Karnaugh suggested that we think of the map as being inscribed on a torus, or donut. It is probably easier just to remember that the table wraps around at the edges. For example, the 1's in the squares corresponding to $\bar{a}\bar{b}\bar{c}$ and $\bar{a}b\bar{c}$ are adjacent, and these two combinations of the three input bits differenly in one bit, b.

We now circle any two 1's that are adjacent horizontally or vertically (not diagonally!) remembering the wrapping rule. It is perfectly fine to have overlapping circles. We also circle 1's that are adjacent to no other 1's, of which there happen to be none in this example (Fig. 8.4).

Because we are guaranteed that adjacent blocks in the grid represent combinations of the three input bits that differ in only one bit (i.e., one of the input bits is 0 in one combination and 1 in the other), we can immediately visually apply the rule: $xy \vee x\bar{y} = x$ (which is really the concensus rule and an absorption law combined). We do this by writing the bits that did not change within the circle for each circle we have drawn, that is, the combination of input bits that

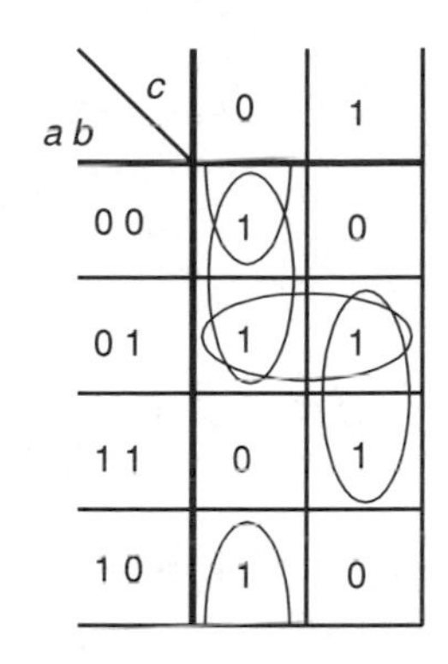

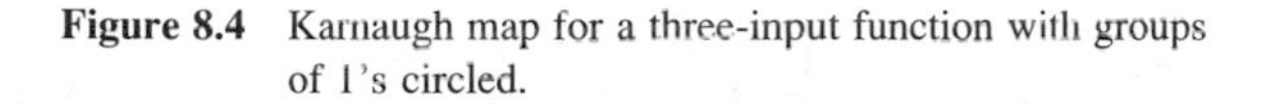
Figure 8.4 Karnaugh map for a three-input function with groups of 1's circled.

exactly characterizes the circle. Thus, we would indicate the circle containing the 1's for the combinations $\bar{a}\bar{b}\bar{c}$ and $\bar{a}b\bar{c}$ by the expression $\bar{a}\ \bar{c}$. Each of the circles represents a prime implicant of the function, and once we write the term corresponding to each circle we merely OR them together to get the BCF of the function:

$$\bar{a}\bar{c} \vee bc \vee \bar{b}\bar{c} \vee \bar{a}b$$

If we are interested in mathematical logic, we want all prime implicants. If we are interested in circuit minimization, however, we do not bother with the prime implicants that are redundant with other prime implicants, i.e., with groupings of 1's in the map in which every 1 overlaps with some other grouping. Thus, in this example, we could represent the same function with fewer logic gates with the expression $a\bar{c} \vee bc \vee \bar{b}\bar{c}$ by eliminating the redundant circle corresponding to $\bar{a}b$.

We may make circles larger than two blocks as long as they neatly eliminate variables. For this to be the case, the circles must contain some power of two number of blocks (i.e., two, four, or eight blocks) and the circled blocks must all be in a straight line (same row or same column) or arranged in a square or rectangle. When we consolidate two adjacent blocks into a circle, we eliminate one of the input bits from the term needed to represent that prime implicant; when we consolidate four blocks, we eliminate two bits; when we consolidate eight blocks, we eliminate three bits.

One of the advantages of Karnaugh maps as a method of minimization is the way in which they take advantage of don't care outputs. If there are don't care bits in the output column of a function's truth table, we may treat them as 0 or 1 when writing the function's Karnaugh map according to whichever choice facilitates creating larger blocks of 1's. For example, consider the four-bit function specified by the truth table ("*d*" indicates *don't care*) shown in Table 8.5.

Because this function has four input bits, its Karnaugh map will have 16 blocks with all possible combinations of the first two input bits along the side and all possible combinations of the other two input bits along the top (all in

TABLE 8.5

a	b	c	d	
0	0	0	0	1
0	0	0	1	d
0	0	1	0	0
0	0	1	1	1
0	1	0	0	0
0	1	0	1	0
0	1	1	0	1
0	1	1	1	0
1	0	0	0	d
1	0	0	1	1
1	0	1	0	1
1	0	1	1	1
1	1	0	0	1
1	1	0	1	d
1	1	1	0	0
1	1	1	1	0

gray code, of course). Once we put 1's and *d*s in the appropriate blocks of the Karnaugh map, we have the grid shown in Figure 8.5.

Now we start circling collections of 1's that consist of two, four, or eight 1's arranged in a straight line, a square, or a rectangle. The wrapping rule makes things a little more complicated, but not much. First, notice that the block corre-

ab \ cd	0 0	0 1	1 1	1 0
0 0	1	d	1	0
0 1	0	0	0	1
1 1	1	d	0	0
1 0	d	1	1	1

Figure 8.5 Karnaugh map of a four-input function with don't care outputs.

sponding to the combination $(a, b, c, d) = (0110)$ has no filled-in neighbors, so it gets its own circle. Thus $\bar{a}bc\bar{d}$ is a prime implicant of the function. If we treat the "d" in the lower left-hand corner of the map as a 1, the bottom row (corresponding to $[a, b] = [1, 0]$) contains nothing but 1's, so we may circle all four of its blocks. Doing so eliminates two of the four input bits from the term that represents this grouping, $a\bar{b}$. If we had not decided to consider the d to be a 1, we would have had to break the three 1's in the bottom row into two smaller groupings, which would have resulted in more terms with more variables in them.

If we consider the d in the block corresponding to $(a, b, c, d) = (1, 1, 0, 1)$ to be a 1 as well, we can circle the square of four blocks in the lower left-hand corner of the grid. Since there are four blocks in this grouping, we eliminated two of the input bits from this term, which is characterized by the expression $a\bar{c}$ (a and c are the only input bits that do not change within this grouping).

The two remaining 1's in the top row of the map do not obviously fit into any larger allowed grouping. However, if we consider the d in the top row of the map to be a 1, because of the wrapping rule, we may group together the middle two blocks in the top row with the middle two blocks in the bottom row to form a "square" representing the term $\bar{b}d$. Similarly, we collect the first two blocks in the top row together with the first two blocks in the bottom row to form another square, $\bar{b}\bar{c}$. As there are no more valid collections of blocks that we can group together, the final BCF realization of this function is given by

$$\bar{a}bc\bar{d} \vee a\bar{b} \vee a\bar{c} \vee \bar{b}d \vee \bar{b}\bar{c}.$$

Its Karnaugh map is shown in Figure 8.6.

ab \ cd	00	01	11	10
00	1	d	1	0
01	0	0	0	1
11	1	d	0	0
10	d	1	1	1

Figure 8.6 Karnaugh map of a four-input function with don't outputs, with groups of 1's (or don't cares) circled.

EXERCISE 8.4

1–4. For exercises 1–4 in exercise set 8.3, put the expressions in BCF again, this time using Karnaugh maps.

5. If we had considered all of the don't care outputs in the last example to be 0 rather than 1, what would the final expression for the simplified expression look like? Write its Karnaugh map and algebraic expression.

Appendix A

Counting In Base 2

As a Power of 2	Base 10	Base 2
2^0	1	1
2^1	2	10
	3	11
2^2	4	100
	5	101
	6	110
	7	111
2^3	8	1000
	9	1001
	10	1010
	11	1011
	12	1100
	13	1101
	14	1110
	15	1111
2^4	16	10000
	17	10001
	18	10010
	19	10011
	20	10100
	21	10101
	22	10110
	23	10111
	24	11000
	25	11001
	26	11010
	27	11010
	28	11010
	29	11010
	30	11010
	31	11010
2^5	32	100000
	33	100001
	34	100010
	35	100011
	36	100100
	37	100101
	38	100110
	39	100111
	40	101000

Appendix B

Powers of 2

2^0	1
2^1	2
2^2	4
2^3	8
2^4	16
2^5	32
2^6	64
2^7	128
2^8	256
2^9	512
2^{10}	1,024
2^{11}	2,048
2^{12}	4,096
2^{13}	8,182
2^{14}	16,384
2^{15}	32,768
2^{16}	65,536
2^{17}	131,072
2^{18}	262,144
2^{19}	524,288
2^{20}	1,048,576
2^{21}	2,097,152
2^{22}	4,194,304
2^{23}	8,388,608
2^{24}	16,777,216
2^{25}	33,554,432
2^{26}	67,108,864
2^{27}	134,217,728
2^{28}	268,435,456
2^{29}	536,870,912
2^{30}	1,073,741,824

Appendix C

Summary of Boolean Functions

AND

common contraction: AND

algebraic symbol: $x \wedge y$ (alternatively, xy)

circuit diagram:

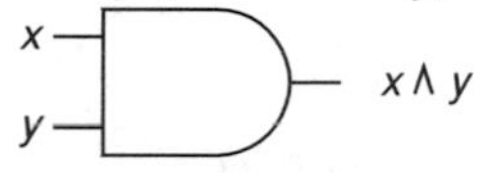

truth table:

x	y	$x \wedge y$
0	0	0
0	1	0
1	0	0
1	1	1

English interpretation: "x is 1 and y is 1."

function in the algebra of sets: intersection

symbol in the algebra of sets: $x \cap y$

Venn diagram:

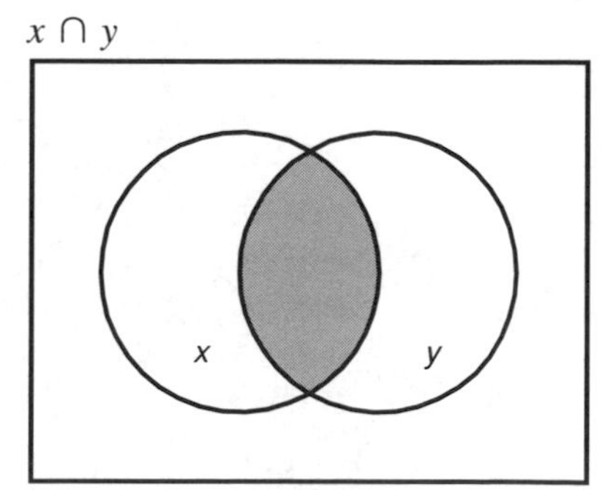

symmetric: yes

OR

common contraction: OR

algebraic symbol: $x \vee y$

circuit diagram:

truth table:

x	y	$x \vee y$
0	0	0
0	1	1
1	0	1
1	1	1

English interpretation: "either x is 1 or y is 1 (or both)."

function in the algebra of sets: union

symbol in the algebra of sets: $x \cup y$

Venn diagram:

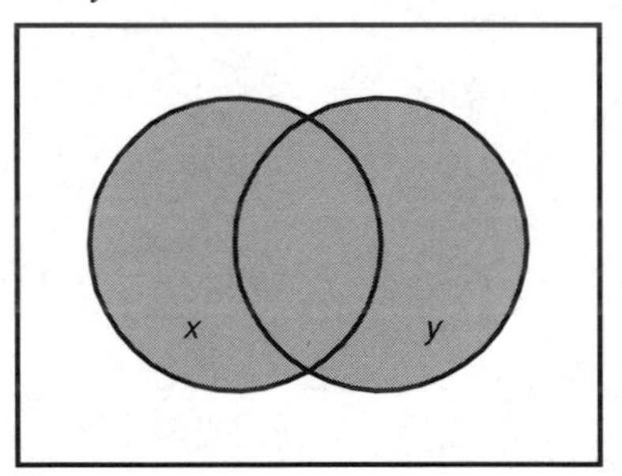

symmetric: yes

NOT

common contraction: NOT

algebraic symbol: $\bar{x}$

circuit diagram:

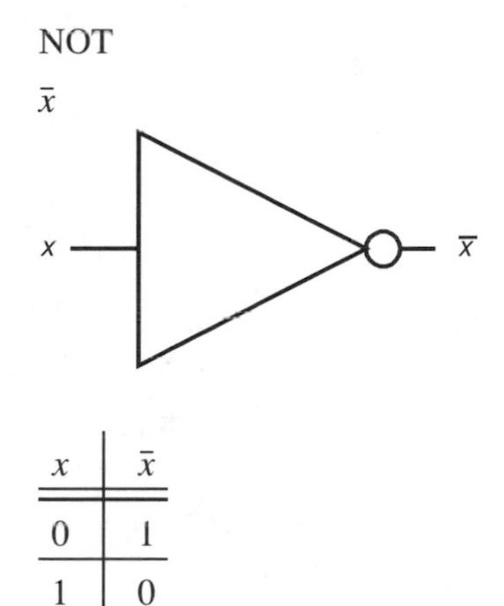

truth table:

x	$\bar{x}$
0	1
1	0

English interpretation: "x is 0."

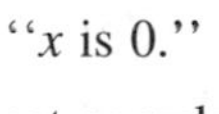

function in the algebra of sets: set complementation

symbol in the algebra of sets: x'

Venn diagram:

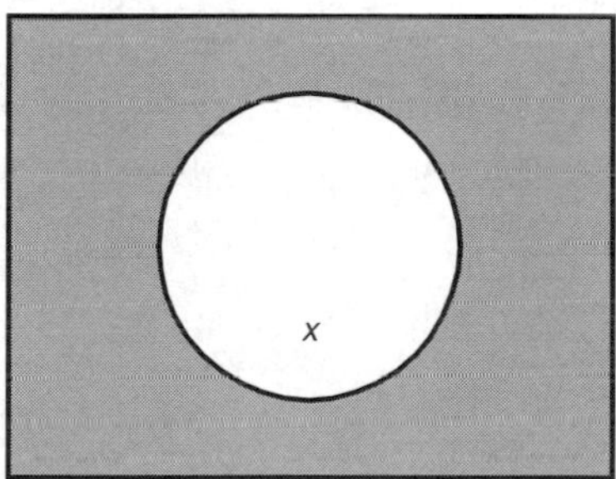

symmetric: not applicable

Exclusive OR

common contraction: XOR

algebraic symbol: $x \oplus y$

circuit diagram:

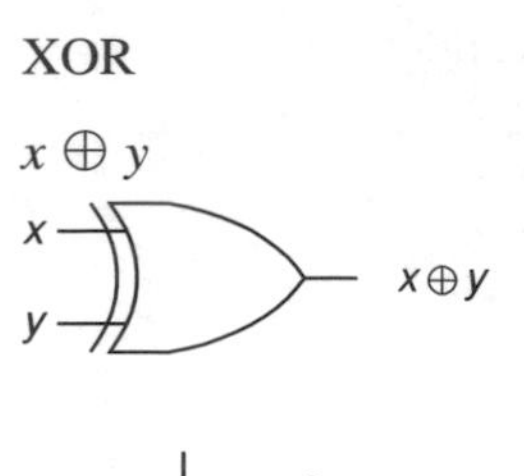

truth table:

x	y	$x \oplus y$
0	0	0
0	1	1
1	0	1
1	1	0

English interpretation: "either x is 1 or y is 1 (but not both)."

function in the algebra of sets: not applicable

symbol in the algebra of sets: not applicable

Venn diagram:

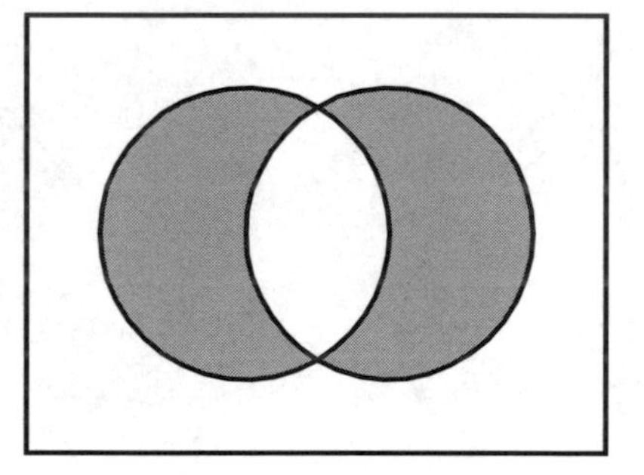

symmetric: yes

Coincidence

common contraction: COIN

algebraic symbol: $x \odot y$

circuit diagrams:

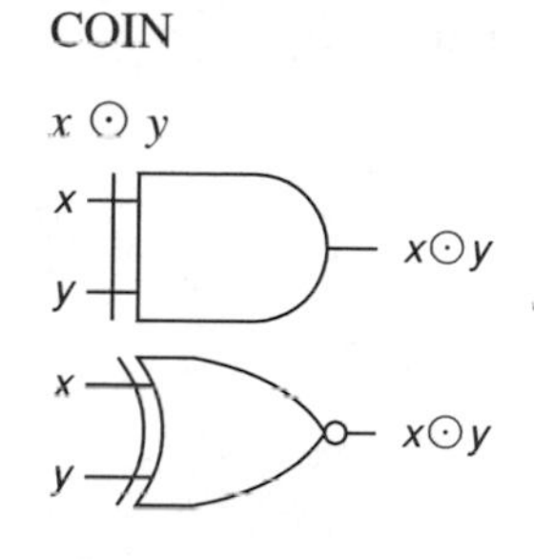

truth table:

x	y	$x \odot y$
0	0	1
0	1	0
1	0	0
1	1	1

English interpretation: "x is equal to y."

function in the algebra of sets: not applicable

symbol in the algebra of sets: not applicable

Venn diagram:

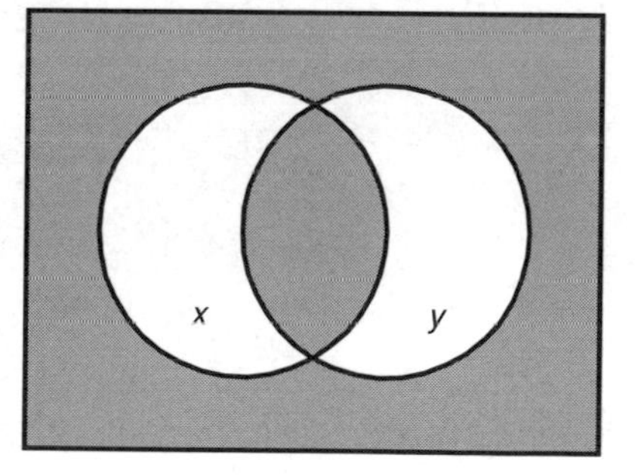

symmetric: yes

NAND

common contraction: NAND

algebraic symbol: $x \uparrow y$

circuit diagram:

x

y

$x \uparrow y$

truth table:

x	y	$x \uparrow y$
0	0	1
0	1	1
1	0	1
1	1	0

English interpretation: "It is not the case that both x and y are 1."

function in the algebra of sets: not applicable

symbol in the algebra of sets: not applicable

Venn diagram:

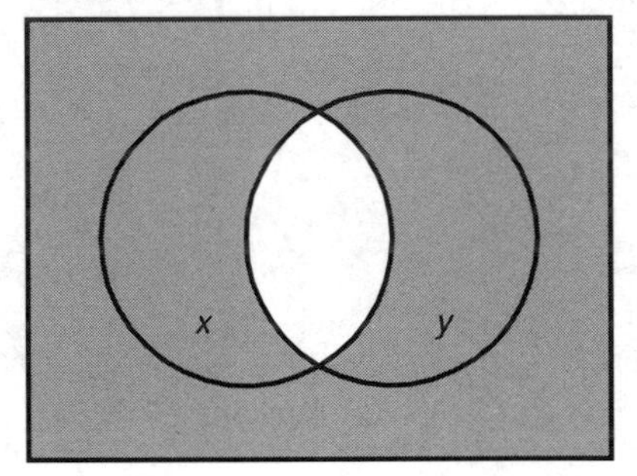

symmetric: yes

NOR

common contraction: NOR

algebraic symbol: $x \downarrow y$

circuit diagram:

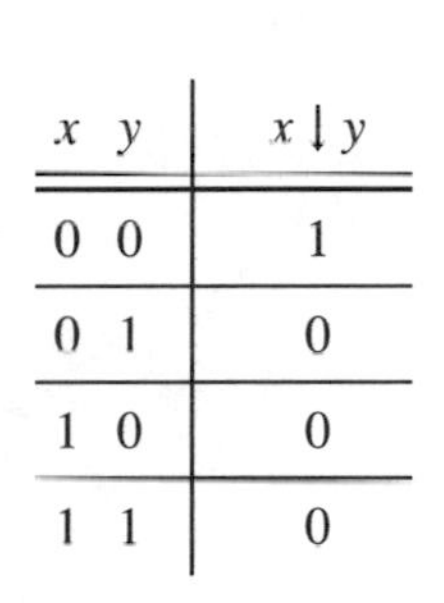

truth table:

x	y	$x \downarrow y$
0	0	1
0	1	0
1	0	0
1	1	0

English interpretation: "It is not the case that either or both of x and y is 1."

function in the algebra of sets: not applicable

symbol in the algebra of sets: not applicable

Venn diagram:

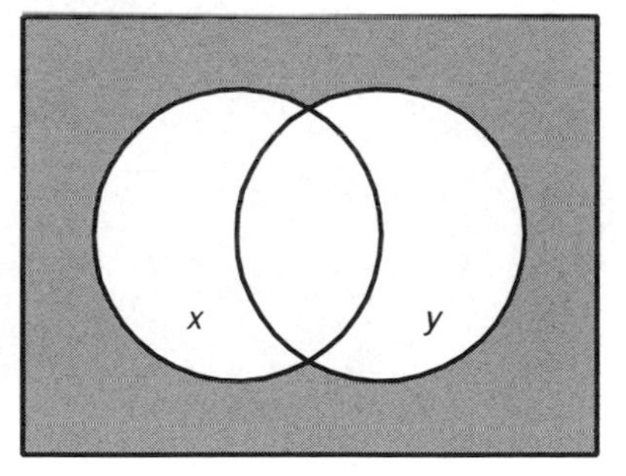

symmetric: yes

Implication

common contraction: Implication

algebraic symbol: $x \rightarrow y$

circuit diagram:

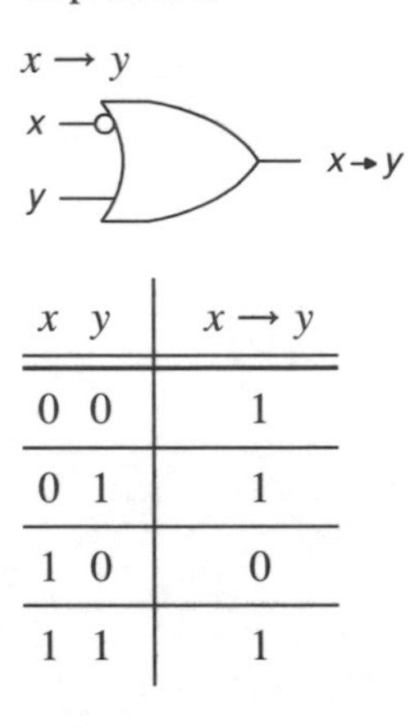

truth table:

x	y	$x \rightarrow y$
0	0	1
0	1	1
1	0	0
1	1	1

English interpretation: "If x is 1, then y is 1."

function in the algebra of sets: subset

symbol in the algebra of sets: $x \subset y$

Venn diagram:

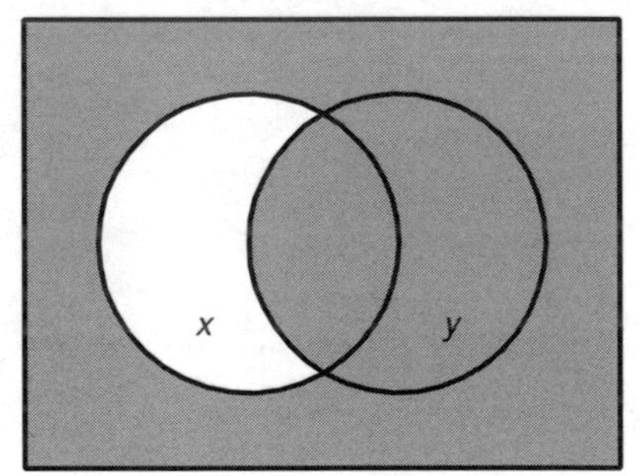

symmetric: no

Further Reading

There are innumerable books about Boolean algebra written from the perspective of digital circuit design and from the perspective of pure mathematics (with regard to which Boolean algebras are one kind among many covered in the field of modern abstract algebra). This list is far from comprehensive, but does provide a good sampling.

DIGITAL ELECTRONICS AND ENGINEERING

I separate one book from the others because while, like them, it is about digital electronics, unlike them it is not an ordinary introductory textbook. It is intended for the layman, and its tone and approach are very much like those of the book you now hold. It contains a lot more physical detail, however, about things like semiconductor fabrication, and different types of transistors.

Maxfield, Clive. *Bebop to the Boolean Boogie.* Solana Beach, CA: Hightext, 1995.

The rest are more or less straightforward introductory texts about digital circuits with a correspondingly hands-on approach to Boolean algebra. Most also contain some coverage of the algebra of sets.

Almaini, A. E. A. *Electronic Logic Systems.* Englewood Cliffs, NJ: Prentice-Hall, 1986.

Booth, Taylor L. *Introduction to Computer Engineering.* 3d ed. New York: John Wiley & Sons, 1984.

Cowan, Sam. *Handbook of Digital Logic . . . With Practical Applications.* Englewood Cliffs, NJ: Prentice-Hall, 1985.

Flegg, Graham. *Boolean Algebra and Its Applications Including Boolean Matrix Algebra.* New York: John Wiley & Sons, 1964.

Friedman, Arthur D. *Fundamentals of Logic Design and Switching Theory.* Rockville, MD: Computer Science Press, 1986.

Hill, Frederick J., and Gerald R. Peterson. *Introduction to Switching Theory and Logical Design.* 3d ed. New York: John Wiley & Sons, 1981.

Hohn, Franz E. *Applied Boolean Algebra.* 2d ed. New York: Macmillan, 1966.

Mange, Daniel. *Analysis and Synthesis of Logic Systems.* Norwood, MA: Artech House, 1986.

Mano, M. Morris. *Computer Logic Design.* Englewood Cliffs, NJ: Prentice-Hall, 1972.

Oberman, R. M. M. *Digital Circuits for Binary Arithmetic.* New York: John Wiley & Sons, 1979.

Sandige, Richard S. *Modern Digital Design.* New York: McGraw-Hill, 1990.

South, G. F. *Boolean Algebra and Its Uses.* London: VNR, 1974.

Strangio, Christopher E. *Digital Electronics: Fundamental Concepts and Applications.* Englewood Cliffs, NJ: Prentice-Hall, 1980.

Unger, Stephen H. *The Essence of Logic Circuits.* 2d ed. New York: IEEE Press, 1989.

ABSTRACT MATHEMATICS AND ALGEBRAS

These are a bit more esoteric, covering Boolean algebra from the more abstract perspective of pure mathematics. They are likely to prove somewhat daunting unless you are comfortable with terms like *rings, modules, fields,* and *semigroups* in a mathematical context.

Hailperin, Theodore. *Boole's Logic and Probability.* Amsterdam: North-Holland, 1976.

Halmos, Paul R. *Algebraic Logic.* New York: Chelsea, 1962.

Halmos, Paul R. *Lectures on Boolean Algebras.* London: VNR, 1963.

Monk, Donald J., and Robert Bonnet, (eds). *Handbook of Boolean Algebras.* Amsterdam: Elsevier, 1989.

Sergiu Rudeanu. *Boolean Functions and Equations.* Amsterdam: North-Holland, 1974.

Sikorski, Roman. *Boolean Algebras.* Berlin: Springer-Verlag, 1964.

HISTORY OF SYMBOLIC LOGIC AND BOOLEAN ALGEBRA

The following selections are histories pertaining to symbolic logic and/or Boolean algebra or original sources. Be prepared for a bewildering variety of notational conventions.

Boole, George. *An Investigation of the Laws of Thought on Which Are Founded the Mathematical Theories of Logic and Probabilities.* New York: Dover, 1958 (originally published 1854).

Carrol, Lewis. *Symbolic Logic.* New York: Potter, 1977.

Couturat, Louis. *The Algebra of Logic.* Chicago: Open Court, 1914.

DeMorgan, Augustus. *On the Syllogism and Other Logical Writings.* New Haven, CT: Yale, 1966.

Karnaugh, M. "The Map Method for Synthesis of Combinational Logic Circuits." *Trans Am IEEE* 72 (1953) 593–599.

Kolmogorov, A. N., and A. P. Yushkevich (eds). *Mathematics of the 19th Century.* Basel: Birkhauswer Verlag, 1992.

Peirce, Charles S. *Writings of Charles S. Peirce.* Bloomington: Indiana University Press, 1982.

Peirce, Charles S. (ed.). *Studies in Logic.* Amsterdam: John Benjamins, 1983 (orig. publ. 1883).

Quine, W. V. *Methods of Logic.* New York: Holt, Rinehart and Winston, 1961.

Quine, W. V. *Selected Logic Papers.* New York: Random House, 1966.

Shannon, Claude E. "A Symbolic Analysis of Relay and Switching Circuits." *Trans Am IEEE* 57 (1938) 713–723. Reprinted in *Claude Elwood Shannon: Collected Papers.* Sloan N. J. A., and Aaron D. Wyner (eds.) New York: IEEE Press, 1992.

Shearman, A. T. *The Development of Symbolic Logic.* Dubuque: Brown Reprint, 1971 (originally published 1906).

Styazhkin, N. I. *History of Mathematical Logic from Leibniz to Peano.* Cambridge: MIT, 1969.

Veitch, E. W. "A Chart Method For Simplifying Truth Functions." Proceedings of the Association for Computing Machinery, Pittsburgh, PA, May 1952.

Venn, John. *Symbolic Logic.* New York: Chelsea, 1971 (orig. publ. 1894).

MATHEMATICAL LOGIC

The following are a few of the countless modern texts about symbolic logic (which includes elements of Boolean logic and the algebra of sets).

Adler, Irving. *Thinking Machines: A Layman's Introduction to Logic, Boolean Algebra, and Computers.* New York: John Day, 1961.

Copi, Irving M. *Symbolic Logic.* 3d ed. New York: Macmillan, 1967.

Dowsing, R. D., V. J. Rayward-Smith, and C. D. Walter. *A First Course in Formal Logic and Its Applications in Computer Science.* Oxford, UK: Blackwell, 1986.

Kalish, Donald, and Richard Montague *Logic: Techniques of Formal Reasoning.* 2d ed. New York: Harcourt Brace Jovanovich, 1980.

Korfhage, Robert R. *Logic and Algorithms with Applications to the Computer and Information Sciences.* New York: John Wiley & Sons, 1966.

Mendelson, Elliott. *Introduction to Mathematical Logic.* 2d ed. New York: Van Nostrand, 1979.

NUMBER SYSTEMS AND GENERAL MATHEMATICS

The following are selections that expand on the themes discussed in chapter 0: numbers and their representations, and the historical development of those systems of representation.

Barrow, John D. *Pi in the Sky: Counting, Thinking, and Being.* Oxford: OUP, 1992.

Bell, E. T. *The Development of Mathematics.* New York: McGraw-Hill, 1940.

Glaser, Anton. *A History of Binary and Other Nondecimal Numeration.* Southampton, PA: Tomash, 1971.

Hardy, G. H. *A Mathematician's Apology.* Cambridge: Cambridge University Press, 1940.

Ifrah, Georges. *From One to Zero: A Universal History of Numbers.* New York: Viking, 1985.

Menninger, Karl. *Number Words and Number Symbols: A Cultural History of Numbers.* Cambridge: MIT, 1969.

HISTORY OF COMPUTERS

The following are books about the history of the development of digital circuits and computers.

Aspray, William (ed). *Computing Before Computers.* Ames, IA: Iowa State University, 1990.

Augarten, Stan. *Bit by Bit: An Illustrated History of Computers.* New York: Ticknor & Fields, 1984.

Gardner, Martin. *Logic Machines and Diagrams.* Chicago: University of Chicago, 1958.

Goldstine, Herman H. *The Computer from Pascal to von Neumann.* Princeton: Princeton University Press, 1972.

Lindgren, Michael. *Glory and Failure: The Difference Engines of Johann Muller, Charles Babbage and Georg and Edvard Scheutz.* (translated by Craig McKay) Linkoping, Sweden: Linkoping University, 1987.

Martin, Ernst. *The Calculating Machines.* (translated and edited by Peggy Kidwell and Michael Williams, originally published in 1925) Cambridge, MA and Los Angeles: MIT Press and Tomash, 1992.

Murray, Francis J. *Mathematical Machines, Volume I: Digital Computers.* New York: Columbia U, 1961.

Randell, Brian (ed). *The Origins of Digital Computers: Selected Papers* (second edition). New York: Springer-Verlag, 1975.

SYLLOGISTIC REASONING

The following selections elaborate on the topics covered in Chapter 8.

Brown, Frank Markham. *Boolean Reasoning: The Logic of Boolean Equations.* Dordrecht: Kluwer, 1990.

Blake, A. ''*Canonical Expressions in Boolean Algebra*'' (dissertation, Dept. of Mathematics, University of Chicago, 1937).

Answers to Exercises

Chapter 0

Exercise 0.1

1. 1 (9^0), 9 (9^1), 81 (9^2), 729 (9^3), 6561 (9^4)
2. 1 (5^0), 5 (5^1), 25 (5^2), 125 (5^3), 625 (5^4)
3. 1 (4^0), 4 (4^1), 16 (4^2), 64 (4^3), 256 (4^4)
4. 1 (3^0), 3 (3^1), 9 (3^2), 27 (3^3), 81 (3^4)
5. 1 (2^0), 2 (2^1), 4 (2^2), 8 (2^3), 16 (2^4)

Exercise 0.2

1. 1, 2, 3, 4, 5, 6, 7, 10, 11, 12, 13, 14, 15, 16, 17, 20, 21, 22, 23, 24
2. 1, 2, 3, 4, 5, 10, 11, 12, 13, 14, 15, 20, 21, 22, 23, 24, 25, 30, 31, 32
3. 1, 2, 3, 4, 10, 11, 12, 13, 14, 20, 21, 22, 23, 24, 30, 31, 32, 33, 34, 40
4. 1, 2, 10, 11, 12, 20, 21, 22, 100, 101, 102, 110, 111, 112, 120, 121, 122, 200, 201, 202
5. 1, 10, 11, 100, 101, 110, 111, 1000, 1001, 1010, 1011, 1100, 1101, 1110, 1111, 10000, 10001, 10010, 10011, 10100

6. 1096
7. 446
8. 10509
9. 500
10. 3666
11. 63
12. 63
13. 63
14. 3123
15. 200
16. 21520
17. 1022200
18. 1000011
19. 134
20. 105
21. 363
22. 185
23. 1000000
24. Given the representation of a number x in base n and base n^p, every p digits in the base n representation can be converted directly to a single base n^p digit. For example, the base 2 representation of 37_{10} is 100101 and the base 8 (2^3) representation is 45. Note that a sequence of three base 2 digits holds exactly the same amount of information that one base 8 (''octal'') digit does: both are capable of representing the base 10 numbers 0 through 7 (0–7 in base 8 and 000–111 in base 2). So we can think of the base 2 number, 100101, as two sequences of three base two digits each, 100 101, that we can convert separately to their octal equivalents, 4 and 5, giving us the base 8 number 45. Similarly, the base 16 number 2B may be thought of as two collections of four base 2 digits, 2 (0010) and B (1011), giving us a base 2 equivalent of 101011 (having dropped the leading zeros). Thus it is easy to convert between base n and base n^p without the intermediate step of converting to base 10. This is why computer programmers often work in base 8 or base 16 instead of the unwieldy base 2.

Exercise 0.3

1. 23 ($14_{10} + 5_{10} = 19_{10}$)
2. 211 ($14_{10} + 42_{10} = 56_{10}$)
3. 1240 ($196_{10} + 116_{10} = 312_{10}$)
4. 1052 ($579_{10} + 197_{10} = 776_{10}$)
5. 1001 ($170_{10} + 47_{10} = 217_{10}$)
6. 1042 ($113_{10} + 129_{10} = 242_{10}$)

Exercise 0.4

1. 10000_5 or 625_{10} (0000_5 through 4444_5)

2. 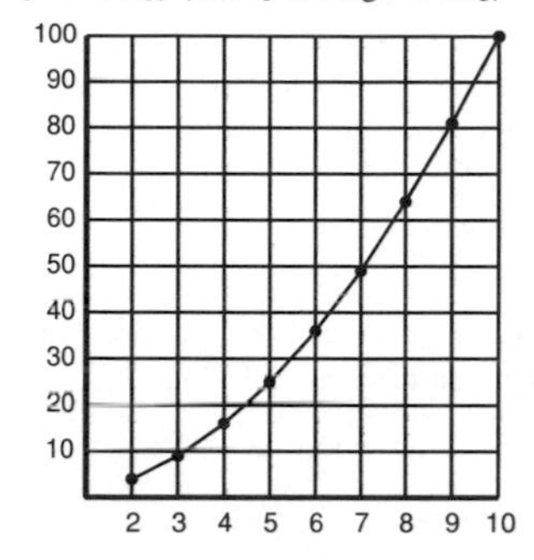

3. Because it makes no sense to speak of a base 0 or base 1 number system; for Hindu-Arabic style number systems to work, there must be a zero digit as well as at least one nonzero digit, so the smallest base we can have is base 2.

4. 3^7 or 2187. Because each of the seven travelers can have one of three drinks, the question is the same as that of how many seven-digit base 3 numbers there are.

5. 3^7 or 4096. The problem is the same as the previous one, except that we have added one more drink option (''none'').

6. 4^{213}

7. When we shift a base 10 number to the left, we multiply it by 10. When we shift it to the right, we divide it by 10, dropping any remainder.

8. $422_5 = 112_{10}$; $4220_5 = 560_{10}$; $42200_5 = 2800_{10}$; $42_5 = 22_{10}$; $4_5 = 4_{10}$. Shifting a base 5 number to the left multiplies it by 5; shifting it to the right divides it by 5, dropping any remainder.

9. $11010010_2 = 210_{10}$; $1101001_2 = 105_{10}$; $110100_2 = 52_{10}$; $11010_2 - 26_{10}$; $1101_2 = 13_{10}$; $110_2 = 6_{10}$; $11_2 = 3_{10}$; $1_2 = 1_{10}$. Each base 10 number is half of the previous one, rounded down (i.e., the remainder is dropped).

10. Shifting a base x number y places to the right divides the number of x^y and shifting it y places to the left multiplies it by x^y.

Exercise 0.5

1. 2^7 or 128. Because there are two ways the coin could fall, the number of combinations of heads and tails in seven flips is the same as the number of seven-digit base 2 numbers. In n flips there are 2^n different combinations.

2. n years from now the substance will be $1/2^n$ as radioactive as it is now (in 1 year it will be $1/2^1$ or 1/2 as radioactive and in 2 years it will be $1/2^2$ or 1/4 as radioactive, and so on). We need to know how many years it will be before the substance is 1/300th as radioactive as it is now (or less, because we are rounding up to the nearest year), so we need to know the smallest n such that 2^n is greater than or equal to 300 (so that $1/2^n$ is less than 1/300th). 2^8 is 256, which is just shy of 300, but 2^9 is 512, so after 9 years the substance will be less than 1/300th as

radioactive as it is now. It will always be somewhat radioactive because there is no finite value of n for which $1/2^n$ is 0.

3. There are 10 individual votes during the show, each of which could turn out thumbs up or thumbs down (i.e., Mike's vote and Arman's vote on each of five movies). Thus there are 2^{10} or 1024 different combinations of ways they could vote, ranging from both Mike and Arman disliking all five movies to both of them liking all five movies. The question is the same as that of how many 10-digit base 2 numbers there are.

4. It would be a point with 2^0 or 1 vertex.

5. If for each option the option may be present or absent in the car, the question is the same as that of how many 37-digit base 2 numbers there are and the answer is 2^{37}.

Chapter 1

Exercise 1.1

1. 1
2. 0
3. 1
4. 0
5. 1
6. 1
7. 0
8. 1
9. 1
10. 0
11. 0
12. 0
13. 1
14. Indeterminate
15. 0
16. Indeterminate

Chapter 2

Exercise 2.1

1.

a)

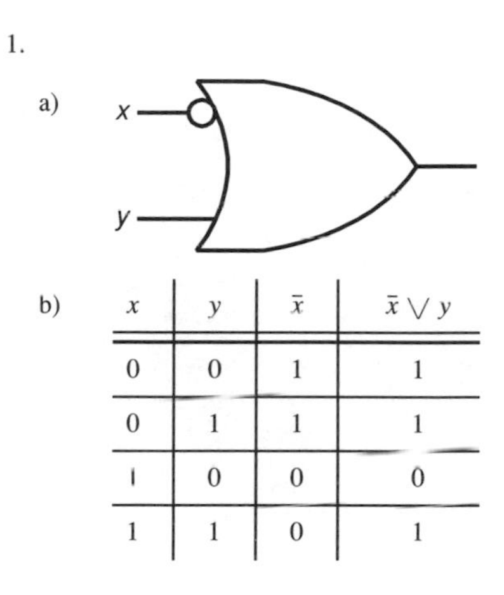

b)

x	y	$\bar{x}$	$\bar{x} \vee y$
0	0	1	1
0	1	1	1
1	0	0	0
1	1	0	1

2.

a)

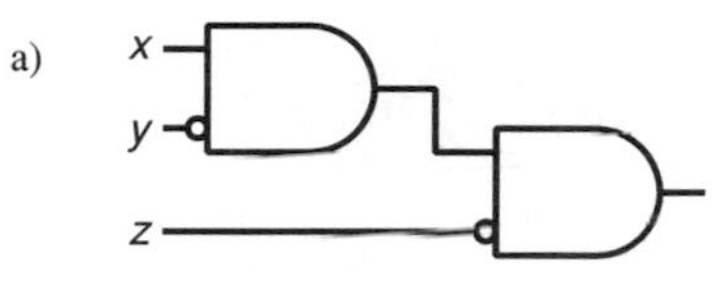

b)

x	y	z	$\bar{y}$	$x \wedge \bar{y}$	$\bar{z}$	$(x \wedge \bar{y}) \wedge \bar{z}$
0	0	0	1	0	1	0
0	0	1	1	0	0	0
0	1	0	0	0	1	0
0	1	1	0	0	0	0
1	0	0	1	1	1	1
1	0	1	1	1	0	0
1	1	0	0	0	1	0
1	1	1	0	0	0	0

3.

a)

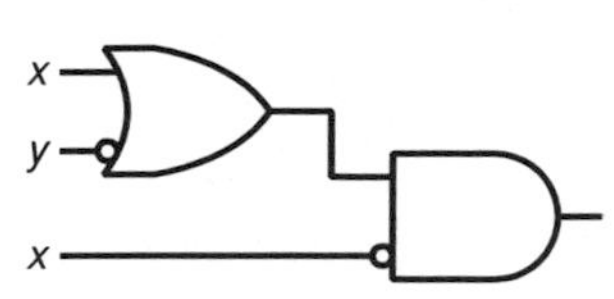

b)

x	y	$\bar{y}$	$x \vee \bar{y}$	$\bar{x}$	$(x \vee \bar{y}) \wedge \bar{x}$
0	0	1	1	1	1
0	1	0	0	1	0
1	0	1	1	0	0
1	1	0	1	0	0

4.

a)

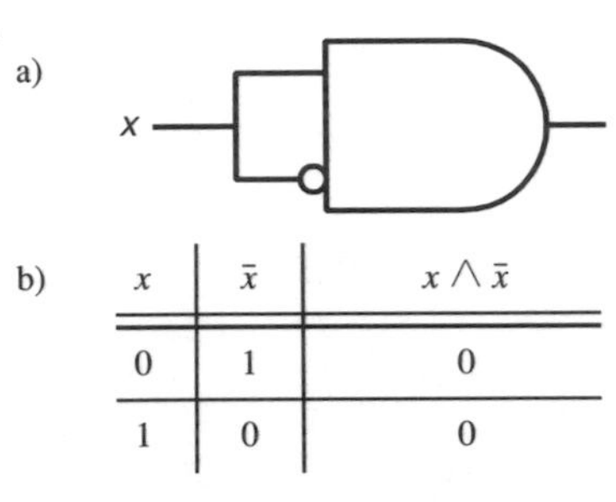

b)

x	$\bar{x}$	$x \wedge \bar{x}$
0	1	0
1	0	0

5.

a)

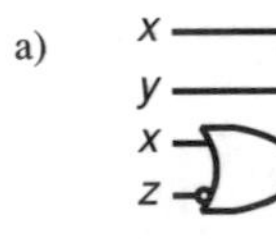

b)

x	y	z	$\bar{z}$	$x \vee \bar{z}$	$y \wedge (x \vee \bar{z})$	$\bar{x}$	$\bar{x} \vee [y \wedge (x \vee \bar{z})]$
0	0	0	1	1	0	1	1
0	0	1	0	0	0	1	1
0	1	0	1	1	1	1	1
0	1	1	0	0	0	1	1
1	0	0	1	1	0	0	0
1	0	1	0	1	0	0	0
1	1	0	1	1	1	0	1
1	1	1	0	1	1	0	1

6.

a)

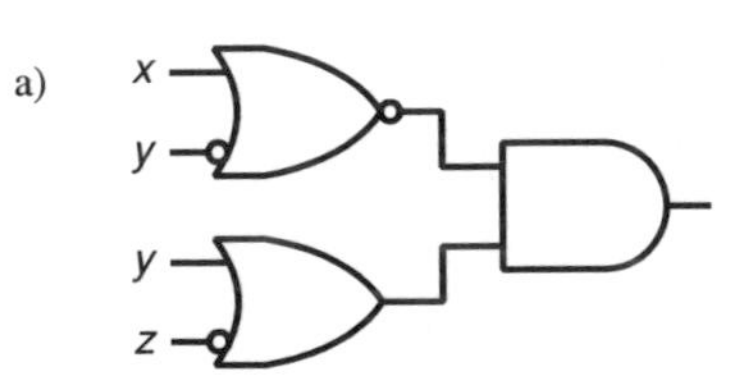

b)

x	y	z	$\bar{y}$	$x \vee \bar{y}$	$(\overline{x \vee \bar{y}})$	$\bar{z}$	$y \vee \bar{z}$	$(\overline{x \vee \bar{y}}) \wedge (y \vee \bar{z})$
0	0	0	1	1	0	1	1	0
0	0	1	1	1	0	0	0	0
0	1	0	0	0	1	1	1	1
0	1	1	0	0	1	0	1	1
1	0	0	1	1	0	1	1	0
1	0	1	1	1	0	0	0	0
1	1	0	0	1	0	1	1	0
1	1	1	0	1	0	0	1	0

7.

a)

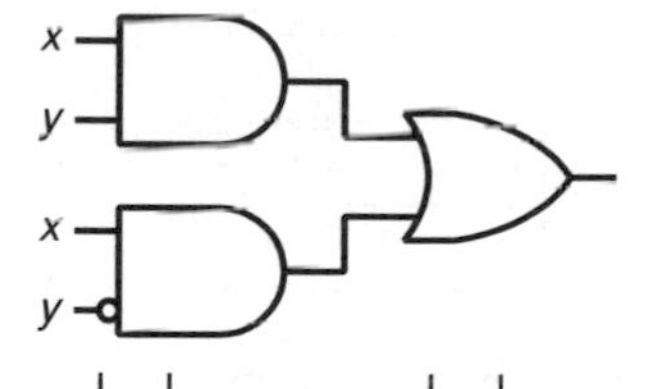

b)

x	y	$x \wedge y$	$\bar{y}$	$x \wedge \bar{y}$	$(x \wedge y) \vee (x \wedge \bar{y})$
0	0	0	1	0	0
0	1	0	0	0	0
1	0	0	1	1	1
1	1	1	0	0	1

8.

a)

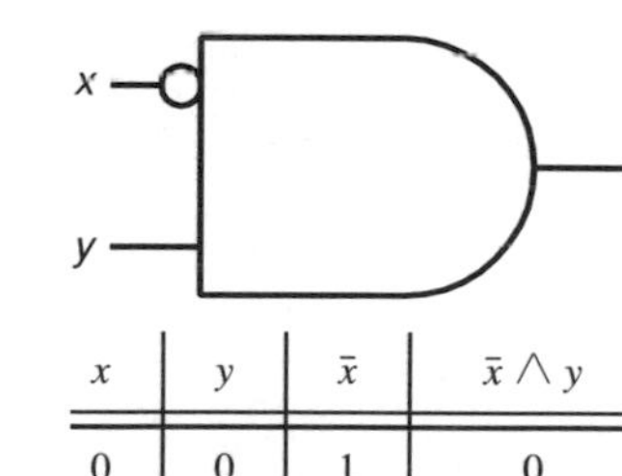

b)

x	y	$\bar{x}$	$\bar{x} \wedge y$
0	0	1	0
0	1	1	1
1	0	0	0
1	1	0	0

9.

a)

b)

a	b	c	d	$\bar{a} \wedge b$	$(\bar{a} \wedge b) \vee c$	$c \vee \bar{d}$	$((\bar{a} \wedge b) \vee c) \wedge (c \vee \bar{d})$
0	0	0	0	0	0	1	0
0	0	0	1	0	0	0	0
0	0	1	0	0	1	1	1
0	0	1	1	0	1	1	1
0	1	0	0	1	1	1	1
0	1	0	1	1	1	0	0
0	1	1	0	1	1	1	1
0	1	1	1	1	1	1	1
1	0	0	0	0	0	1	0
1	0	0	1	0	0	0	0
1	0	1	0	0	1	1	1
1	0	1	1	0	1	1	1
1	1	0	0	0	0	1	0
1	1	0	1	0	0	0	0
1	1	1	0	0	1	1	1
1	1	1	1	0	1	1	1

10.

a)

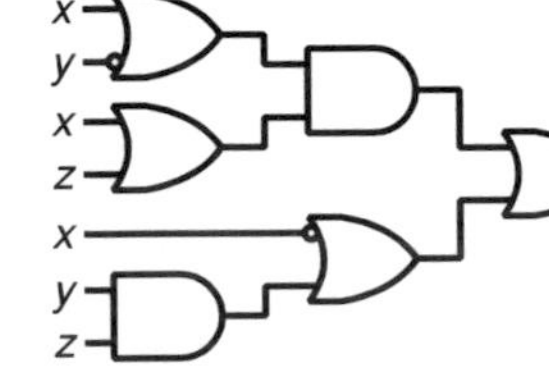

b)

x	y	z	$(x \vee \bar{y}) \wedge (x \vee z)$	$\bar{x} \vee (y \wedge z)$	$[(x \vee \bar{y}) \wedge (x \vee z)] \vee [\bar{x} \vee (y \wedge z)]$
0	0	0	0	1	1
0	0	1	1	1	1
0	1	0	0	1	1
0	1	1	0	1	1
1	0	0	1	0	1
1	0	1	1	0	1
1	1	0	1	0	1
1	1	1	1	1	1

This function is always 1. Its inputs have no actual effect on the output.

11.

a)

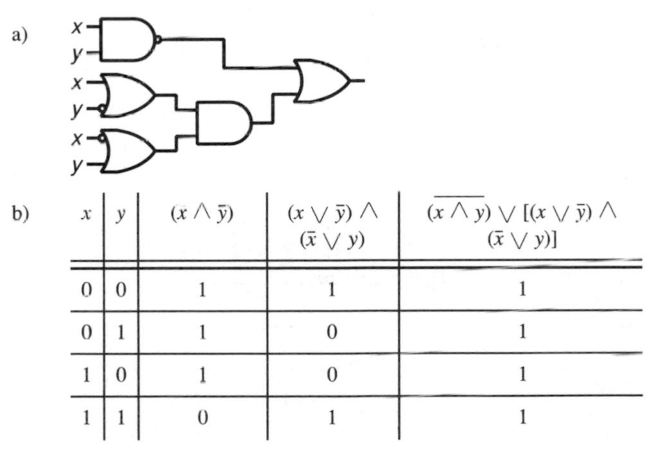

b)

x	y	$(x \wedge \bar{y})$	$(x \vee \bar{y}) \wedge (\bar{x} \vee y)$	$\overline{(x \wedge y)} \vee [(x \vee \bar{y}) \wedge (\bar{x} \vee y)]$
0	0	1	1	1
0	1	1	0	1
1	0	1	0	1
1	1	0	1	1

12.

a) $(\bar{x} \wedge \bar{y}) \wedge y$

b)

x	y	$\bar{x} \wedge \bar{y}$	$(\bar{x} \wedge \bar{y}) \wedge y$
0	0	1	0
0	1	0	0
1	0	0	0
1	1	0	0

13.

a) $(\bar{x} \vee y) \wedge (\bar{y} \wedge z)$

b)

x	y	z	$x \vee y$	$\bar{y} \wedge z$	$(x \vee y) \wedge (\bar{y} \wedge z)$
0	0	0	0	0	0
0	0	1	0	1	0
0	1	0	1	0	0
0	1	1	1	0	0
1	0	0	1	0	0
1	0	1	1	1	1
1	1	0	1	0	0
1	1	1	1	0	0

14.

a) $x \vee \bar{x}$

b)

x	$\bar{x}$	$x \vee \bar{x}$
0	1	1
1	0	1

15.

a) $x \vee (y \vee z)$

b)

x	y	z	$y \vee z$	$x \vee (y \vee z)$
0	0	0	0	0
0	0	1	1	1
0	1	0	1	1
0	1	1	1	1
1	0	0	0	1
1	0	1	1	1
1	1	0	1	1
1	1	1	1	1

16.

a) $\overline{(x \vee \bar{y})} \vee (x \wedge z)$

b)

x	y	z	$x \vee y$	$\overline{(x \vee y)}$	$x \wedge z$	$\overline{(x \vee y)} \vee (x \wedge z)$
0	0	0	0	1	0	1
0	0	1	0	1	0	1
0	1	0	1	0	0	0
0	1	1	1	0	0	0
1	0	0	1	0	0	0
1	0	1	1	0	1	1
1	1	0	1	0	0	0
1	1	1	1	0	1	1

17.

a) $[a \vee \overline{(b \wedge c)}] \wedge b$

b)

a	b	c	$b \wedge c$	$\overline{(b \wedge c)}$	$[a \vee \overline{(b \wedge c)}]$	$[a \vee \overline{(b \wedge c)}] \wedge b$
0	0	0	0	1	1	0
0	0	1	0	1	1	0
0	1	0	0	1	1	1
0	1	1	1	0	0	0
1	0	0	0	1	1	0
1	0	1	0	1	1	0
1	1	0	0	1	1	1
1	1	1	1	0	1	1

18.

a) $(a \vee \bar{c}) \wedge b$

b)

a	b	c	$\bar{c}$	$a \vee \bar{c}$	$(a \vee \bar{c}) \wedge b$
0	0	0	1	1	0
0	0	1	0	0	0
0	1	0	1	1	1
0	1	1	0	0	0
1	0	0	1	1	0
1	0	1	0	1	0
1	1	0	1	1	1
1	1	1	0	1	1

The functions in Exercises 17 and 18 yield the same bit in all eight rows of their respective truth tables. They are, therefore, equivalent functions. When written down they differ in that the $(b \wedge c)$ in Exercise 17 is replaced with $\bar{c}$ in Exercise 18. Although replacing one with the other still yields the same function in the end, they themselves are not equal in all cases. Their differences, however, are cancelled out by the operations of the other components of the function.

19.

a) $(\bar{x} \wedge z) \vee (y \wedge \bar{z})$

b)

x	y	z	$\bar{x}$	$\bar{x} \wedge z$	$\bar{z}$	$y \wedge \bar{z}$	$(\bar{x} \wedge z) \vee (y \wedge \bar{z})$
0	0	0	1	0	1	0	0
0	0	1	1	1	0	0	1
0	1	0	1	0	1	1	1
0	1	1	1	1	0	0	1
1	0	0	0	0	1	0	0
1	0	1	0	0	0	0	0
1	1	0	0	0	1	1	1
1	1	1	0	0	0	0	0

20. $\bar{x} \wedge y$

21. x

22. $(\bar{x} \wedge \bar{y}) \vee (x \wedge y)$ or alternatively $(x \vee \bar{y}) \wedge (\bar{x} \vee y)$

Exercise 2.2

1.

a)

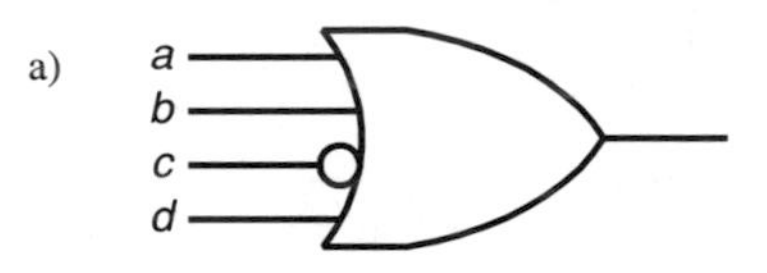

b)

a	b	c	d	$\bar{c}$	$a \vee b \vee \bar{c} \vee d$
0	0	0	0	1	1
0	0	0	1	1	1
0	0	1	0	0	0
0	0	1	1	0	1
0	1	0	0	1	1
0	1	0	1	1	1
0	1	1	0	0	1
0	1	1	1	0	1
1	0	0	0	1	1
1	0	0	1	1	1
1	0	1	0	0	1
1	0	1	1	0	1
1	1	0	0	1	1
1	1	0	1	1	1
1	1	1	0	0	1
1	1	1	1	0	1

2.

a)

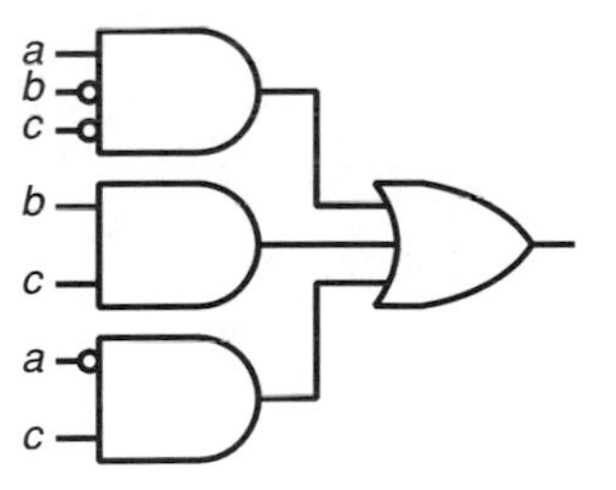

b)

a	b	c	$a \wedge \bar{b} \wedge \bar{c}$	$b \wedge c$	$\bar{a} \wedge c$	$(a \wedge \bar{b} \wedge \bar{c}) \vee (b \wedge c) \vee (\bar{a} \wedge c)$
0	0	0	0	0	0	0
0	0	1	0	0	1	1
0	1	0	0	0	0	0
0	1	1	0	1	1	1
1	0	0	1	0	0	1
1	0	1	0	0	0	0
1	1	0	0	0	0	0
1	1	1	0	1	0	1

3.

a)

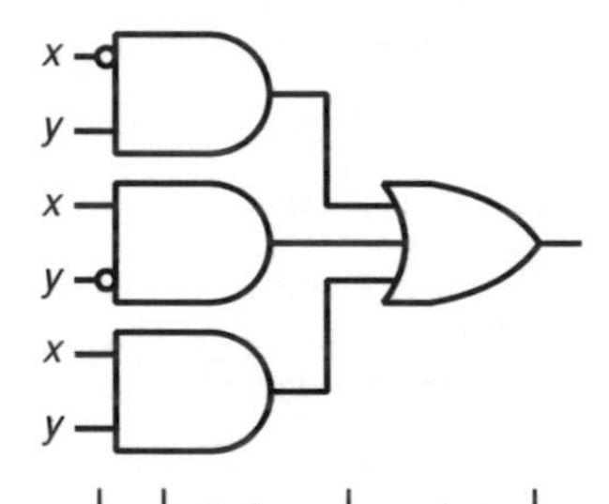

b)

x	y	$(\bar{x} \wedge y)$	$(x \wedge \bar{y})$	$(x \wedge y)$	$(\bar{x} \wedge y) \vee (x \wedge \bar{y}) \vee (x \wedge y)$
0	0	0	0	0	0
0	1	1	0	0	1
1	0	0	1	0	1
1	1	0	0	1	1

4.

a)

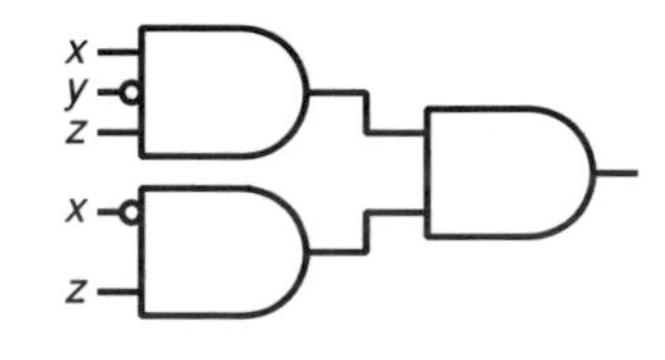

b)

x	y	z	$x \wedge \bar{y} \wedge z$	$\bar{x} \wedge z$	$(x \wedge \bar{y} \wedge z) \wedge (\bar{x} \wedge z)$
0	0	0	0	0	0
0	0	1	0	1	0
0	1	0	0	0	0
0	1	1	0	1	0
1	0	0	0	0	0
1	0	1	1	0	0
1	1	0	0	0	0
1	1	1	0	0	0

5.

a)

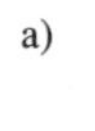

b)

a	b	c	$a \vee \bar{b} \vee c$	$\overline{(a \vee \bar{b} \vee c)}$
0	0	0	1	0
0	0	1	1	0
0	1	0	0	1
0	1	1	1	0
1	0	0	1	0
1	0	1	1	0
1	1	0	1	0
1	1	1	1	0

6.

a)

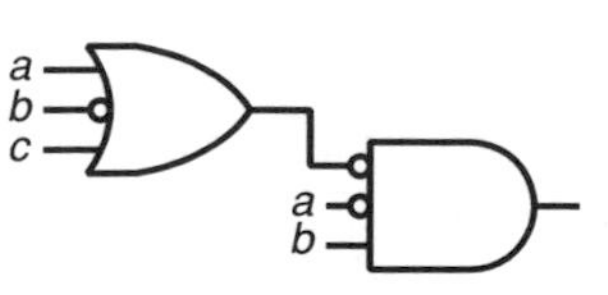

b)

a	b	c	$a \vee \bar{b} \vee c$	$\overline{(a \vee \bar{b} \vee c)}$	$\overline{(a \vee \bar{b} \vee c)} \wedge \bar{a} \wedge b$
0	0	0	1	0	0
0	0	1	1	0	0
0	1	0	0	1	1
0	1	1	1	0	0
1	0	0	1	0	0
1	0	1	1	0	0
1	1	0	1	0	0
1	1	1	1	0	0

7.

a) $(x \wedge \bar{y} \wedge \bar{z}) \vee y$

b)

x	y	z	$x \wedge \bar{y} \wedge \bar{z}$	$(x \wedge \bar{y} \wedge \bar{z}) \vee y$
0	0	0	0	0
0	0	1	0	0
0	1	0	0	1
0	1	1	0	1
1	0	0	1	1
1	0	1	0	0
1	1	0	0	1
1	1	1	0	1

8.

a) $\overline{(x \wedge \bar{y} \wedge \bar{z})}$

b)

x	y	z	$x \wedge \bar{y} \wedge \bar{z}$	$\overline{(x \wedge \bar{y} \wedge \bar{z})}$
0	0	0	0	1
0	0	1	0	1
0	1	0	0	1
0	1	1	0	1
1	0	0	1	0
1	0	1	0	1
1	1	0	0	1
1	1	1	0	1

9.

a) $\overline{(x \wedge \bar{y} \wedge \bar{z})} \vee y$

b)

x	y	z	$x \wedge \bar{y} \wedge \bar{z}$	$\overline{(x \wedge \bar{y} \wedge \bar{z})}$	$\overline{(x \wedge \bar{y} \wedge \bar{z})} \vee y$
0	0	0	0	1	1
0	0	1	0	1	1
0	1	0	0	1	1
0	1	1	0	1	1
1	0	0	1	0	0
1	0	1	0	1	1
1	1	0	0	1	1
1	1	1	0	1	1

The functions in Exercises 8 and 9 differ only in that the function in Exercise 9 has y ORed in at the end. The output columns of the two functions are identical. Each consists of all 1s except a single 0 in the space corresponding to $(x, y, z) = (1, 0, 0)$. In this case y is 0, so ORing y into the function in Exercise 8 will not change the 0 to a 1. On the other hand, ORing in a new variable will never turn any of the 1s into 0s, so ORing the y into the function of Exercise 8 has no effect whatsoever on the output.

10.

a) $\overline{(\bar{a} \vee b \vee \bar{c})} \wedge a \wedge \bar{c}$

b)

a	b	c	$\bar{a} \vee b \vee \bar{c}$	$\overline{(\bar{a} \vee b \vee \bar{c})}$	$\overline{(\bar{a} \vee b \vee \bar{c})} \wedge a \wedge \bar{c}$
0	0	0	1	0	0
0	0	1	1	0	0
0	1	0	1	0	0
0	1	1	1	0	0
1	0	0	1	0	0
1	0	1	0	1	0
1	1	0	1	0	0
1	1	1	1	0	0

11.

a) $(\bar{a} \vee b) \wedge (a \vee \bar{b}) \wedge (a \vee b)$

b)

a	b	$\bar{a} \vee b$	$a \vee \bar{b}$	$a \vee b$	$(\bar{a} \vee b) \wedge (a \vee \bar{b}) \wedge (a \vee b)$
0	0	1	1	0	0
0	1	1	0	1	0
1	0	0	1	1	0
1	1	1	1	1	1

Chapter 3

Exercise 3.1

1.

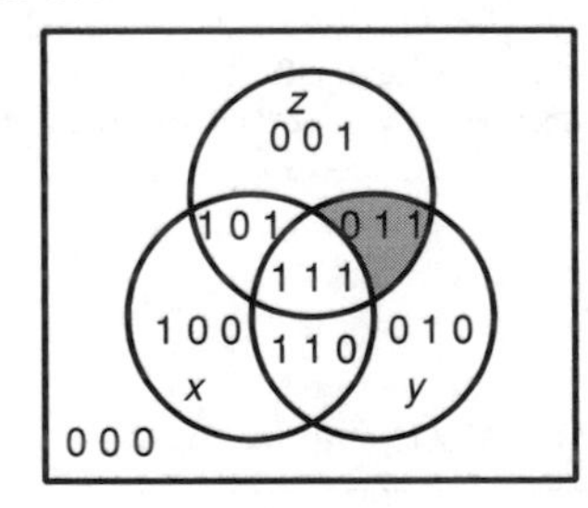

2. The drawing you are most likely to draw looks like this:

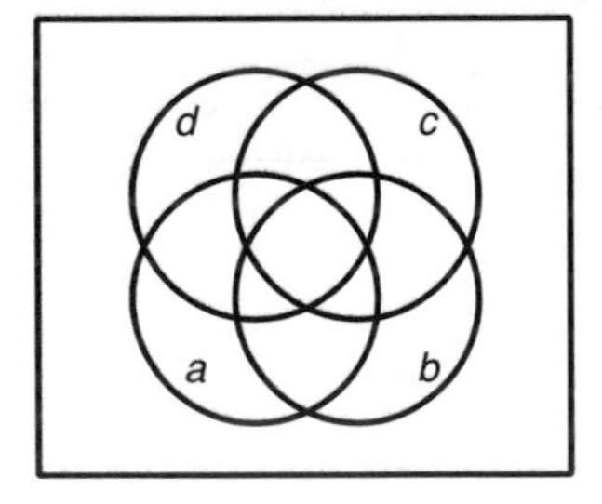

This diagram violates the rule that each new set must bisect every existing closed area in the diagram, thereby doubling the number of individual closed areas in the entire diagram. Specifically, there is no place on the diagram where the two diagonal sets intersect without also including parts of the other sets in the diagram. There are no regions in the diagram that correspond to $(a, b, c, d) = (1, 0, 1, 0)$ or $(a, b, c, d) = (0, 1, 0, 1)$ (regions numbered 5 and 10). If you count, you will find that there are, in fact, only 14 distinct areas in the Venn diagram, not 16 (2^4). John Venn himself, however, drew a valid four-set diagram. His looked more or less like this:

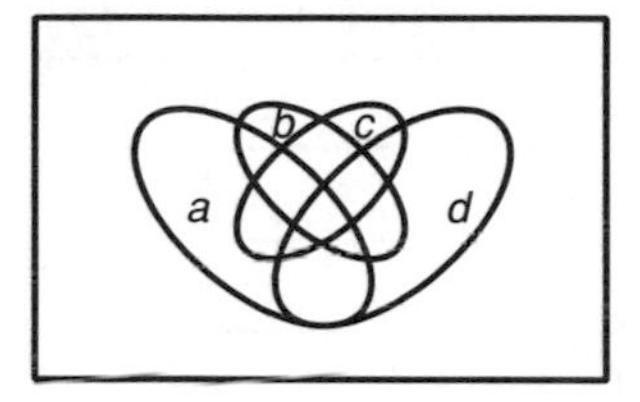

Exercise 3.2

1.

a)

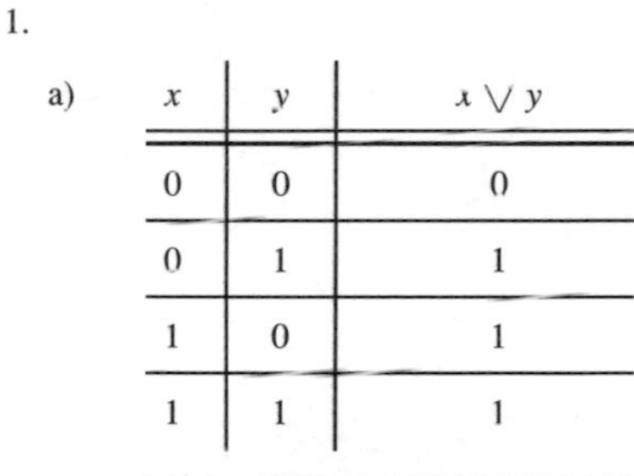

x	y	$x \lor y$
0	0	0
0	1	1
1	0	1
1	1	1

b)

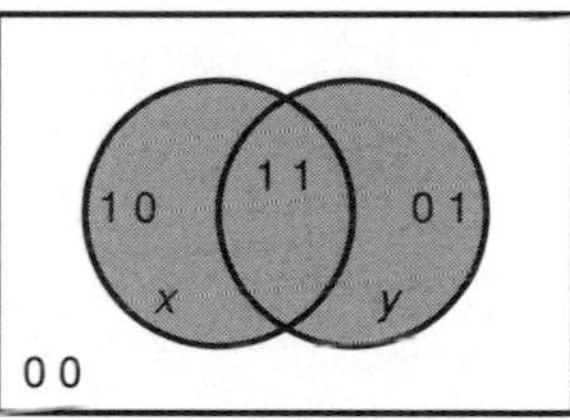

2.

a)

x	y	z	$(x \vee \bar{y} \vee \bar{z}) \wedge (\overline{\bar{x} \wedge y \wedge \bar{z}})$
0	0	0	1
0	0	1	1
0	1	0	0
0	1	1	0
1	0	0	1
1	0	1	1
1	1	0	1
1	1	1	1

b)

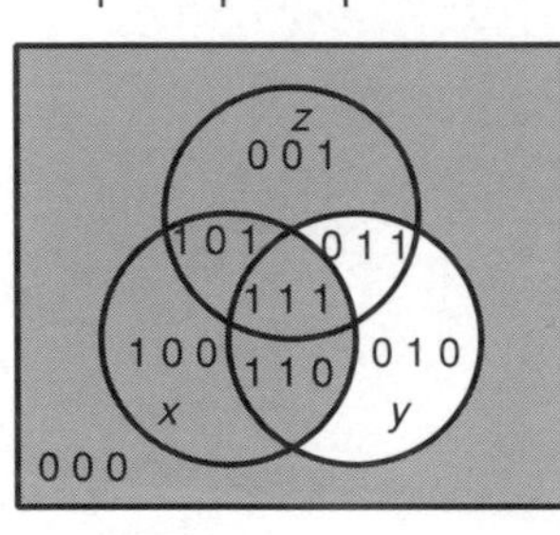

3.

a)

x	y	$\bar{x} \vee y$
0	0	1
0	1	1
1	0	0
1	1	1

b)

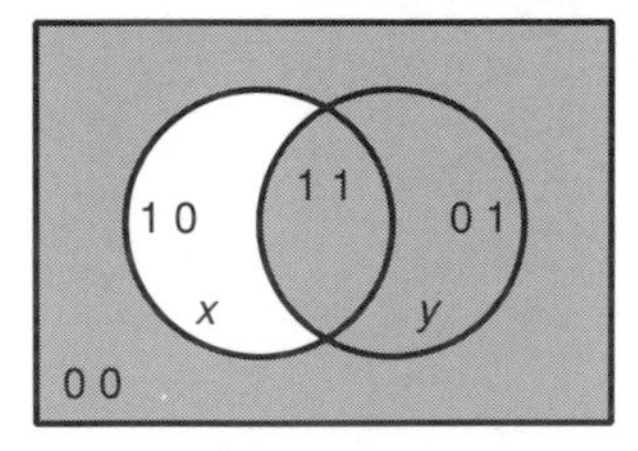

4.

a)

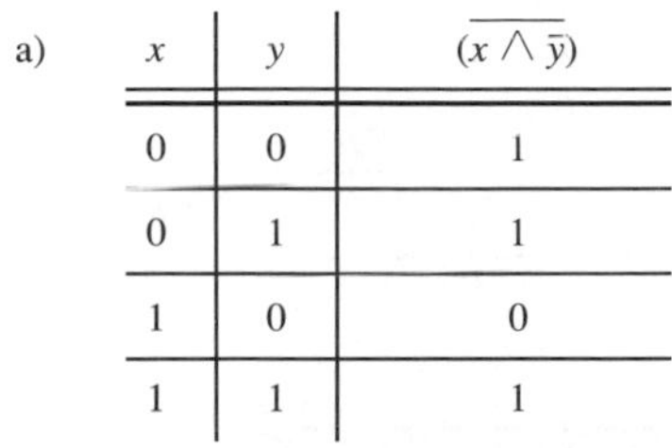

x	y	$(\overline{x \wedge \bar{y}})$
0	0	1
0	1	1
1	0	0
1	1	1

b)

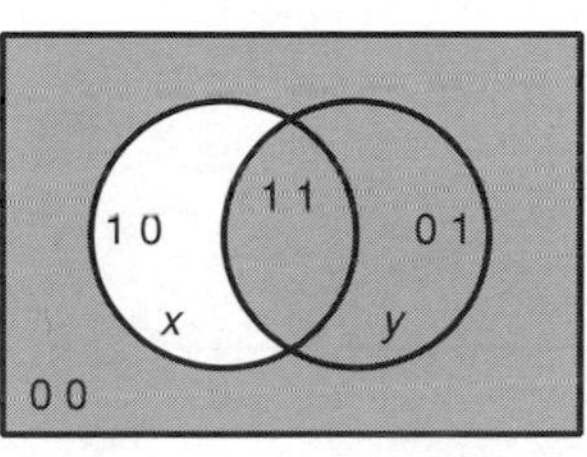

5.

a)

x	y	z	$(x \vee y \vee z) \wedge (x \vee \bar{y} \vee \bar{z}) \wedge$ $(\bar{x} \vee y \vee \bar{z}) \wedge (\bar{x} \vee \bar{y} \vee z)$
0	0	0	0
0	0	1	1
0	1	0	1
0	1	1	0
1	0	0	1
1	0	1	0
1	1	0	0
1	1	1	1

b)

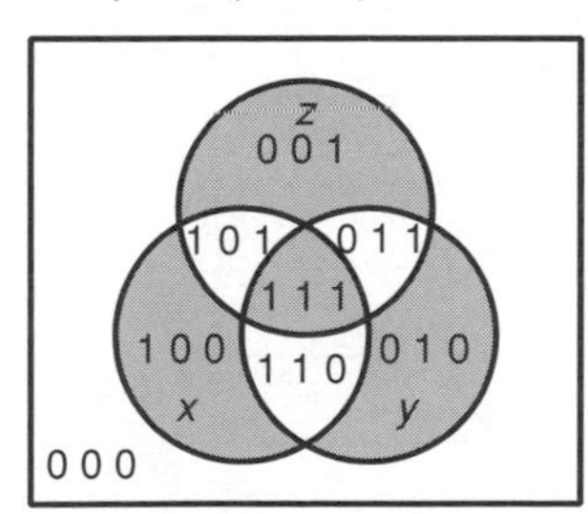

6.

a)

a	b	c	$(\bar{a} \wedge \bar{b} \wedge c) \vee (\bar{a} \wedge b \wedge \bar{c}) \vee (a \wedge \bar{b} \wedge \bar{c}) \vee (a \wedge b \wedge c)$
0	0	0	0
0	0	1	1
0	1	0	1
0	1	1	0
1	0	0	1
1	0	1	0
1	1	0	0
1	1	1	1

b)

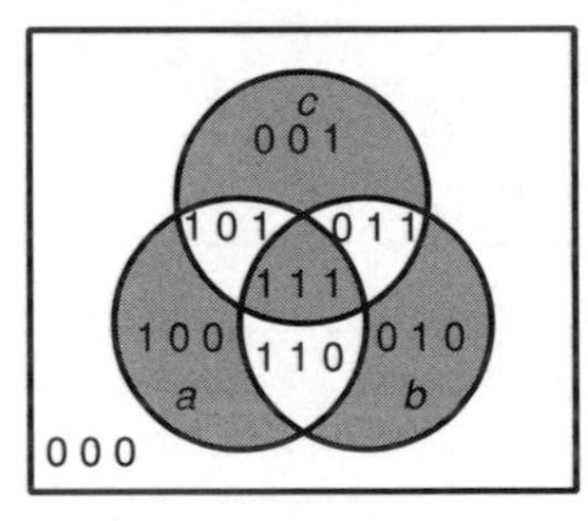

7.

a)

x	y	z	$(x \vee y \vee z) \wedge (x \vee y \vee \bar{z}) \wedge \overline{(x \wedge \bar{y})}$
0	0	0	0
0	0	1	0
0	1	0	1
0	1	1	1
1	0	0	0
1	0	1	0
1	1	0	1
1	1	1	1

b)

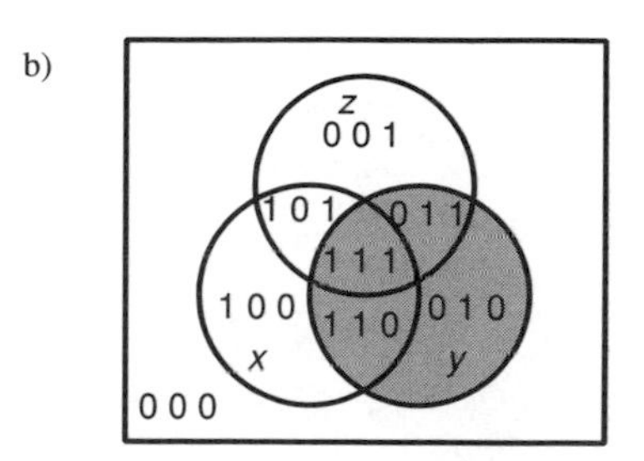

8.

a)

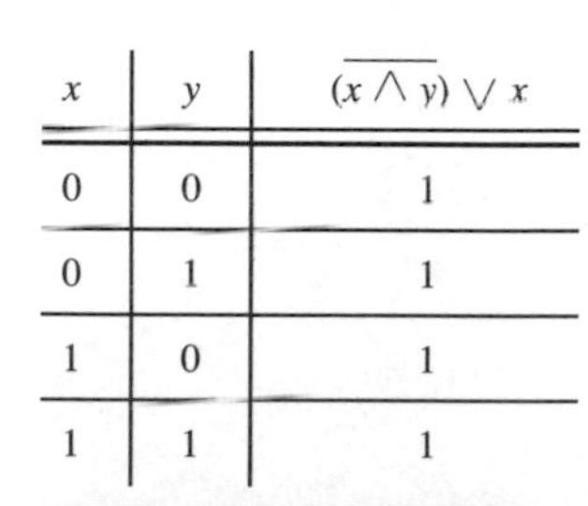

x	y	$(\overline{x \wedge y}) \vee x$
0	0	1
0	1	1
1	0	1
1	1	1

b)

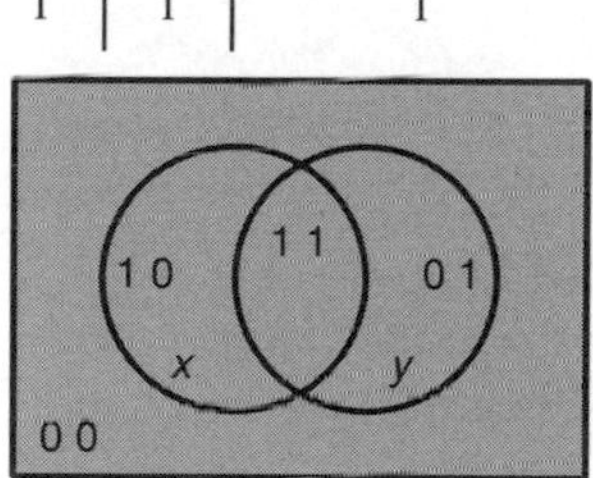

Chapter 4

Exercise 4.1

1.

a)

b)

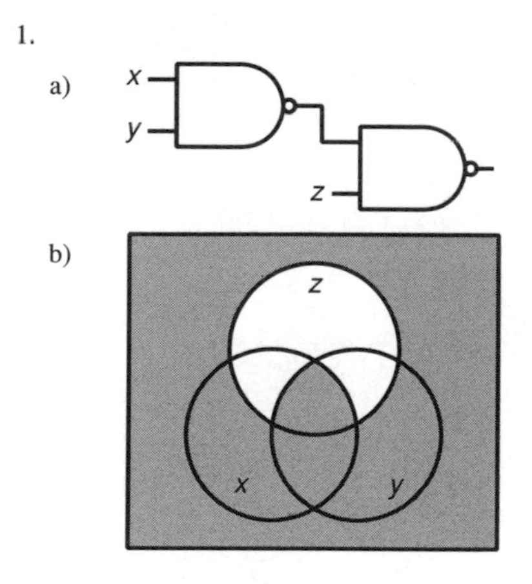

c)

x	y	z	$(x \uparrow y) \uparrow z$
0	0	0	1
0	0	1	0
0	1	0	1
0	1	1	0
1	0	0	1
1	0	1	0
1	1	0	1
1	1	1	1

d) $\overline{[(\overline{x \wedge y}) \wedge z]}$

2.

a)

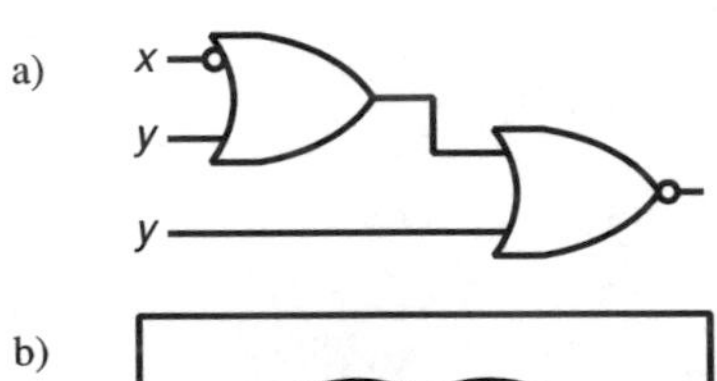

b)

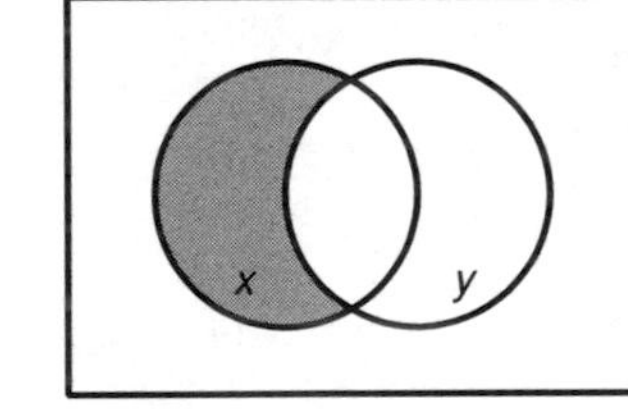

c)

x	y	$(x \rightarrow y) \downarrow y$
0	0	0
0	1	0
1	0	1
1	1	0

d) $\overline{[(\bar{x} \vee y) \vee y]}$

3.

a)

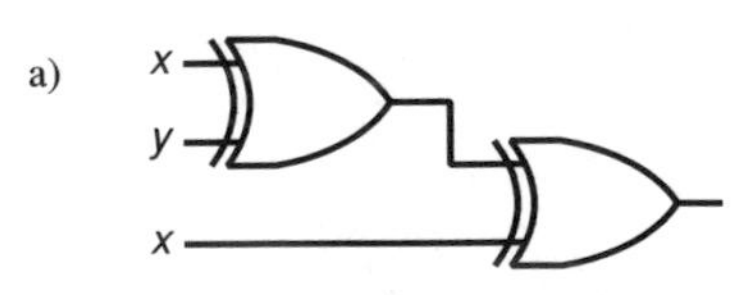

b)

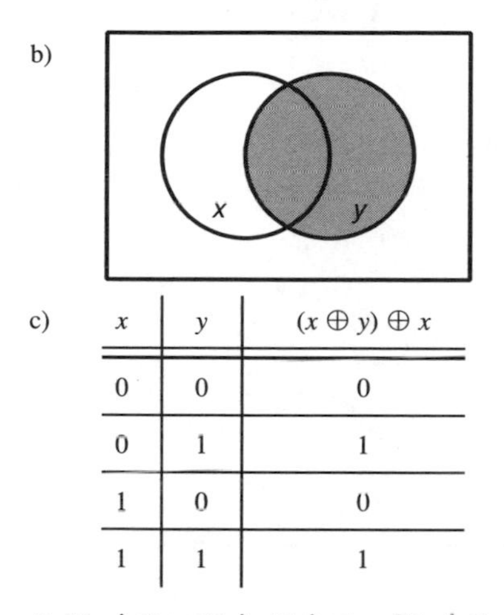

c)

x	y	$(x \oplus y) \oplus x$
0	0	0
0	1	1
1	0	0
1	1	1

d) $[((x \wedge \bar{y}) \vee (\bar{x} \wedge y)] \wedge \bar{x}) \vee [((x \wedge \bar{y}) \vee (\bar{x} \wedge y)) \wedge x]$

4.

a)

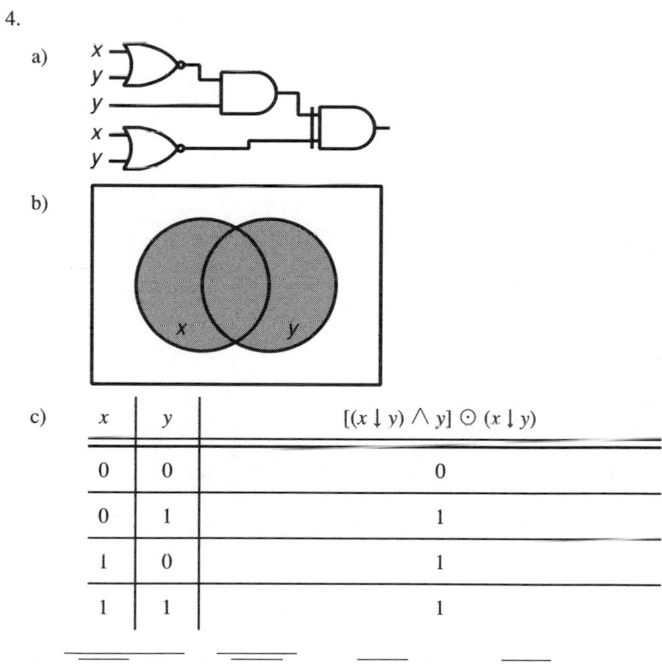

b)

c)

x	y	$[(x \downarrow y) \wedge y] \odot (x \downarrow y)$
0	0	0
0	1	1
1	0	1
1	1	1

d) $[\overline{((\overline{x \vee y}) \wedge y)} \wedge \overline{((\overline{x \vee y}))}] \vee [((\overline{x \vee y}) \wedge y) \wedge (\overline{x \vee y})]$

5.

a)

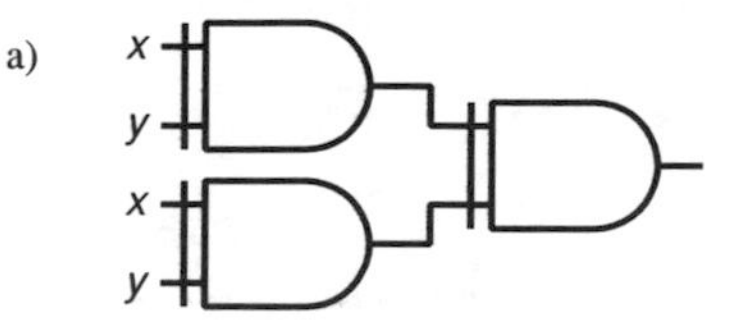

b)

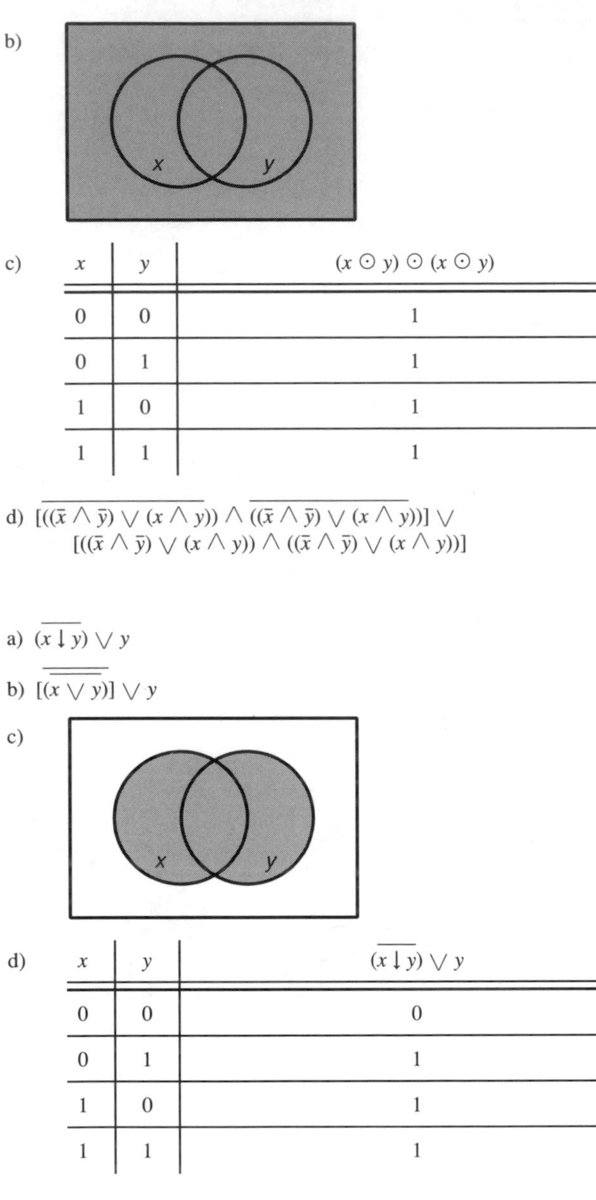

c)

x	y	$(x \odot y) \odot (x \odot y)$
0	0	1
0	1	1
1	0	1
1	1	1

d) $[\overline{((\bar{x} \wedge \bar{y}) \vee (x \wedge y)}) \wedge \overline{((\bar{x} \wedge \bar{y}) \vee (x \wedge y))}] \vee$
$[((\bar{x} \wedge \bar{y}) \vee (x \wedge y)) \wedge ((\bar{x} \wedge \bar{y}) \vee (x \wedge y))]$

6.

a) $\overline{(x \downarrow y)} \vee y$

b) $[\overline{\overline{(x \vee y)}}] \vee y$

c)

d)

x	y	$\overline{(x \downarrow y)} \vee y$
0	0	0
0	1	1
1	0	1
1	1	1

e) $x \vee y$

7.

a) $(x \oplus y) \oplus (x \oplus y)$

b) $[((x \wedge \bar{y}) \vee (\bar{x} \wedge y)) \wedge \overline{((x \wedge \bar{y}) \vee (\bar{x} \wedge y))}] \vee$
$[\overline{((x \wedge \bar{y}) \vee (\bar{x} \wedge y))} \wedge ((x \wedge \bar{y}) \vee (\bar{x} \wedge y))]$

c)

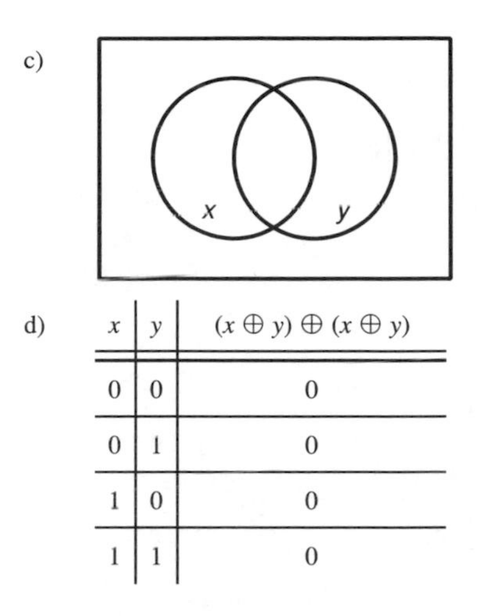

d)

x	y	$(x \oplus y) \oplus (x \oplus y)$
0	0	0
0	1	0
1	0	0
1	1	0

e) 0

8.

a) $(x \uparrow y) \odot (x \oplus y)$

b) $[(\overline{\overline{(x \wedge y)}}) \wedge ((x \wedge \bar{z}) \vee \overline{(\bar{x} \wedge z))]} \vee ((x \wedge y) \wedge ((x \wedge \bar{z}) \vee (\bar{x} \wedge z)))$

c)

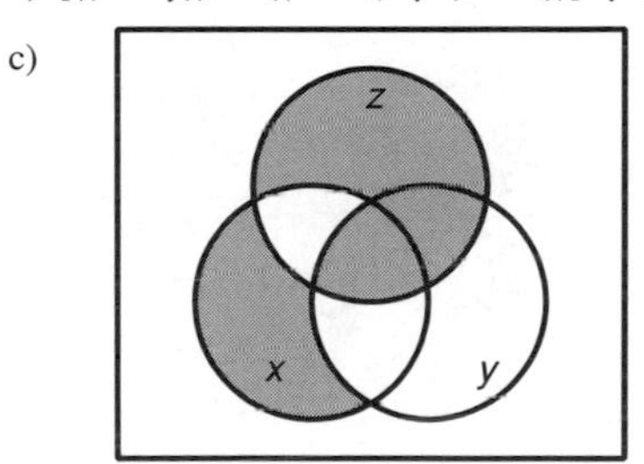

d)

x	y	z	$(x \uparrow y) \odot (x \oplus z)$
0	0	0	0
0	0	1	1
0	1	0	0
0	1	1	1
1	0	0	1
1	0	1	0
1	1	0	0
1	1	1	1

e) There is no simpler realization

9.

a) $(x \odot y) \uparrow (x \vee y)$

b) $\overline{[((\bar{x} \wedge \bar{y}) \vee (x \wedge y)) \wedge (x \vee y)]}$

c)

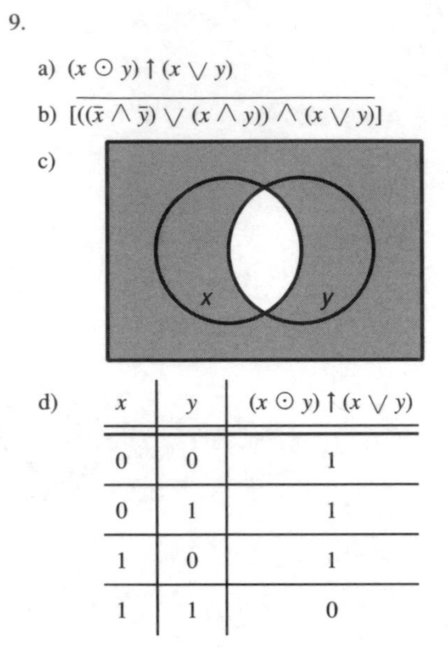

d)

x	y	$(x \odot y) \uparrow (x \vee y)$
0	0	1
0	1	1
1	0	1
1	1	0

e) $x \uparrow y$

10.

a) $(x \rightarrow y) \oplus [(x \downarrow y) \wedge (x \odot y)]$

b) $[(\overline{\bar{x} \vee y}) \wedge ((\overline{x \vee y}) \wedge ((\bar{x} \wedge \bar{y}) \vee (x \wedge y)))] \vee$

$[(\bar{x} \vee y) \wedge \overline{((\overline{x \vee y}) \wedge ((\bar{x} \wedge \bar{y}) \vee (x \wedge y)))}]$

c)

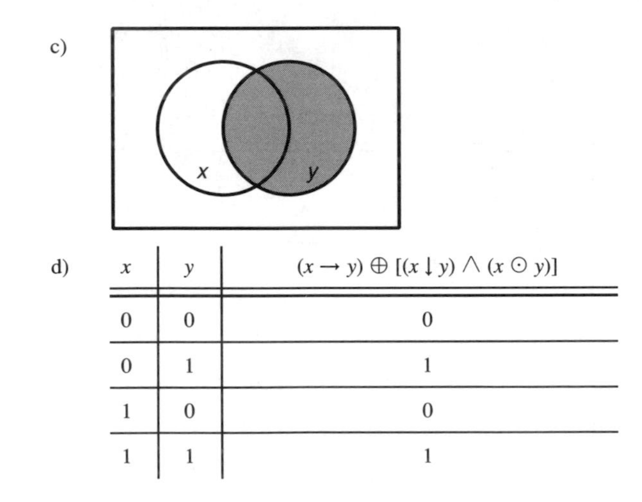

d)

x	y	$(x \rightarrow y) \oplus [(x \downarrow y) \wedge (x \odot y)]$
0	0	0
0	1	1
1	0	0
1	1	1

e) y

11.

Column 10: y
Column 12: x
Column 5: $\bar{y}$
Column 3: $\bar{x}$
Column 4: $x \wedge \bar{y}$
Column 13: $x \vee \bar{y}$

12. The truth table for an n-input bit function will have 2^n rows. There are 2 to the power of that number different ways of filling in the output column then. Thus there are 2^{2^n} distinct Boolean functions of n-input bits.

13. Because the order of the input bits does not matter in a symmetric function, the only things that characterize a symmetric function are (a) the total number of input bits and (b) the number of those input bits that must be 1 to make the function 1. For instance, both AND and OR are symmetric (let us speak only of the 2-input varieties for the moment). Of the 2-input bits, AND is 1 when 2 of the inputs are 1, and 0 when 1 or 0 of them are. OR is 1 when 1 or 2 of the 2-input bits are 1, but 0 when 0 of them are.

There are 2^{n+1} different n-bit symmetric Boolean functions. Consider, for example, all 4-bit symmetric functions. One of them is 1 only when 2 of the inputs are 1. Another is 1 when 0, 3, or all 4 of the inputs are 1, but 0 when only 1 or 2 of the inputs are 1. So let us look at the number of inputs that could be 1: (0, 1, 2, 3, and 4). Now let us assign a bit to each of these possibilities. There are as many different 4-bit symmetric functions as there are combinations of these 5 ($n + 1$) bits. These functions range from the first (which is 0 when 0, 1, 2, 3, or all 4 of the input bits are 1) to the last (which is 1 when 0, 1, 2, 3, or all 4 of the input bits are 1).

Exercise 4.2

1. $x \uparrow y = \overline{(x \wedge y)} = (x \wedge y) \oplus 1$

$x \downarrow y = \overline{(x \vee y)} = [((x \wedge y) \oplus y) \oplus x] \oplus 1$

$x \odot y = \overline{(x \oplus y)} = (x \oplus y) \oplus 1$

2. $x \downarrow y = \overline{(x \vee y)} = (x \vee y) \uparrow (x \vee y) = (\bar{x} \uparrow \bar{y}) \uparrow (\bar{x} \uparrow \bar{y}) = [(x \uparrow x) \uparrow (y \uparrow y)] \uparrow [(x \uparrow x) \uparrow (y \uparrow y)]$

$x \oplus y = (\bar{x} \wedge y) \vee (x \wedge \bar{y}) = [(x \uparrow x) \wedge y] \vee [x \wedge (y \uparrow y)] =$

$\overline{[(x \uparrow x) \uparrow y]} \vee \overline{[x \uparrow (y \uparrow y)]} =$

$\overline{\overline{[((x \uparrow x) \uparrow y)]}} \uparrow \overline{\overline{[(x \uparrow (y \uparrow y))]}} =$

$[(x \uparrow x) \uparrow y] \uparrow [x \uparrow (y \uparrow y)]$

$x \odot y = (\bar{x} \wedge \bar{y}) \vee (x \wedge y) = [(x \uparrow x) \wedge (y \uparrow y)] \vee (x \wedge y) =$

$\overline{[(x \uparrow x) \uparrow (y \uparrow y)]} \vee \overline{(x \uparrow y)} =$

$\overline{\overline{[((x \uparrow x) \uparrow (y \uparrow y))]}} \uparrow \overline{\overline{((x \uparrow y))}} =$

$[(x \uparrow x) \uparrow (y \uparrow y)] \uparrow (x \uparrow y)$

3. a)

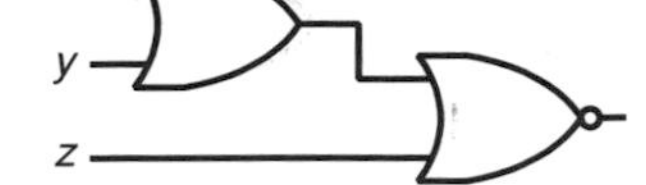

b)

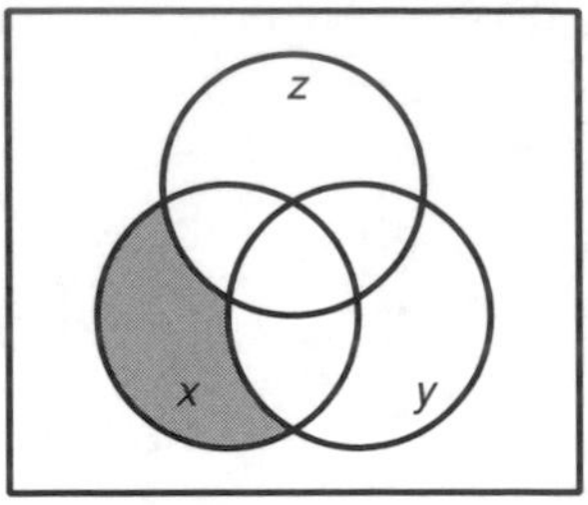

c)

x	y	z	$(x \rightarrow y) \downarrow z$
0	0	0	0
0	0	1	0
0	1	0	0
0	1	1	0
1	0	0	1
1	0	1	0
1	1	0	0
1	1	1	0

d) $\overline{[(\bar{x} \vee y) \vee z]}$ (or alternatively $x \wedge \bar{y} \wedge \bar{z}$)

e)

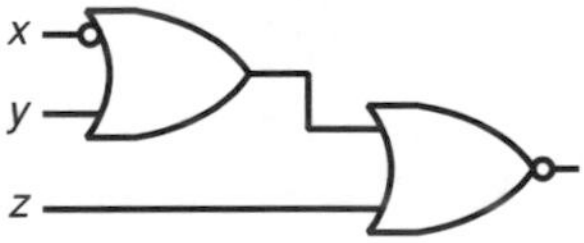

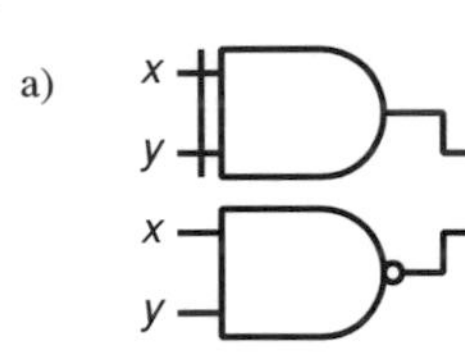

4.

a)

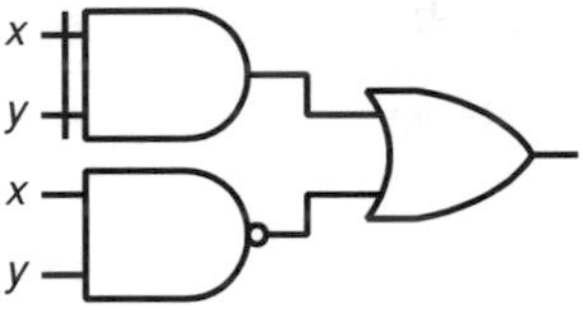

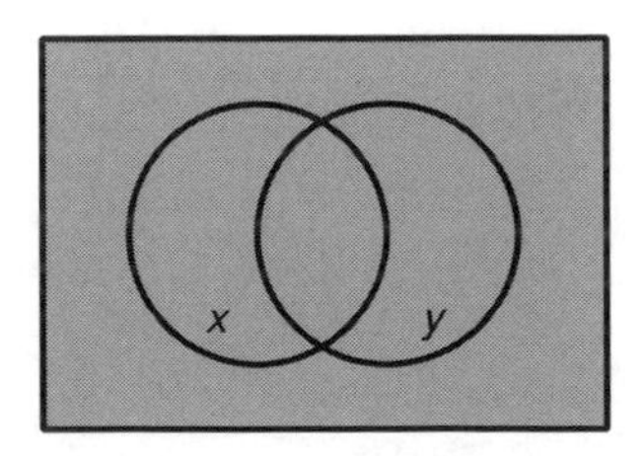

b)

c) 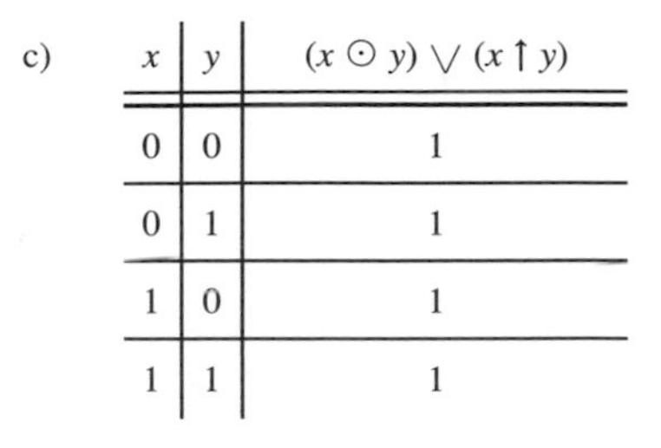

x	y	$(x \odot y) \vee (x \uparrow y)$
0	0	1
0	1	1
1	0	1
1	1	1

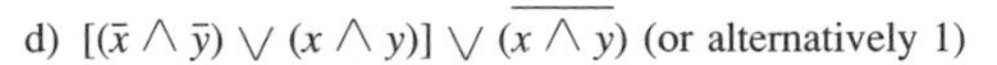

d) $[(\bar{x} \wedge \bar{y}) \vee (x \wedge y)] \vee (\overline{x \wedge y})$ (or alternatively 1)

e) 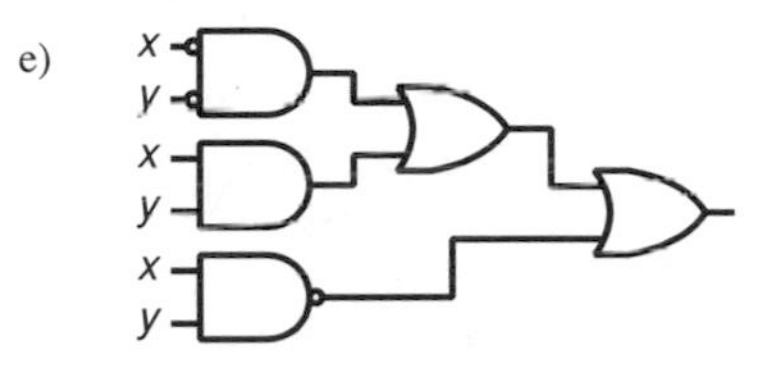

5.

a) 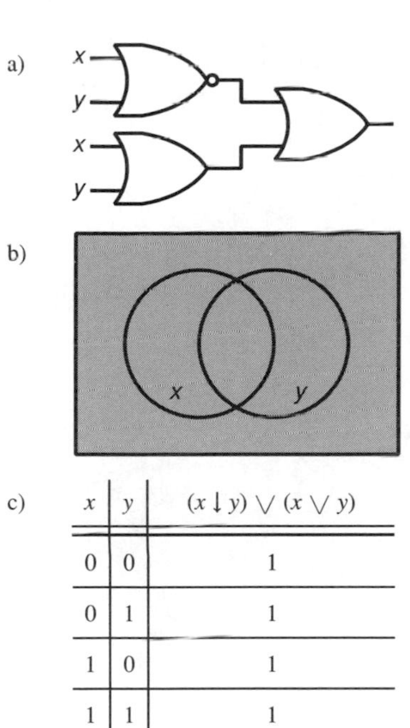

b)

c)

x	y	$(x \downarrow y) \vee (x \vee y)$
0	0	1
0	1	1
1	0	1
1	1	1

d) $(\overline{x \vee y}) \vee (x \vee y)$ (or alternatively 1)

e) 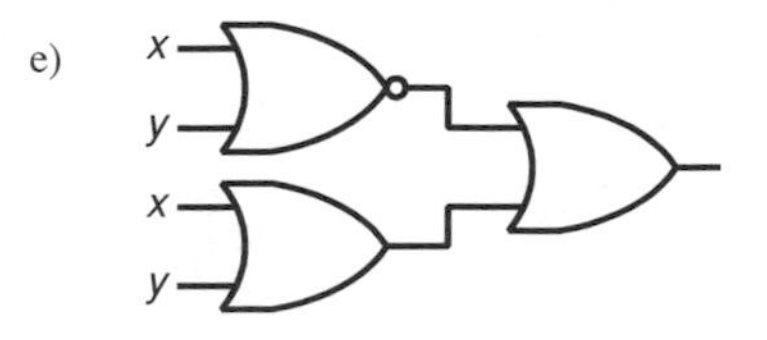

6.

a)

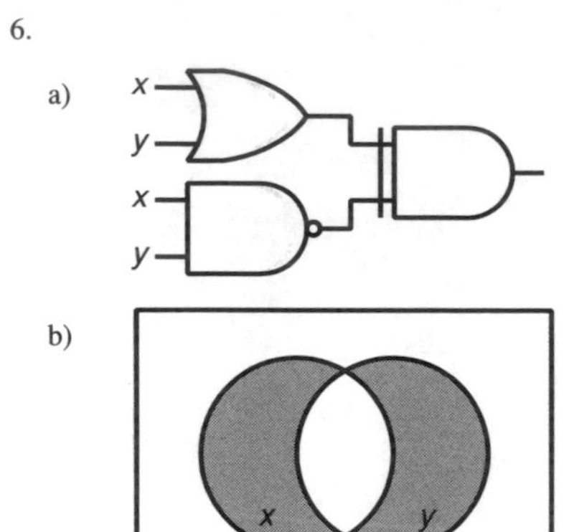

b)

c)

x	y	$(x \vee y) \odot (x \uparrow y)$
0	0	0
0	1	1
1	0	1
1	1	0

d) $[\overline{(x \vee y)} \wedge (x \wedge y)] \vee [(x \vee y) \wedge \overline{(x \wedge y)}]$ (note that this is the same as x $\odot$ y)

e)

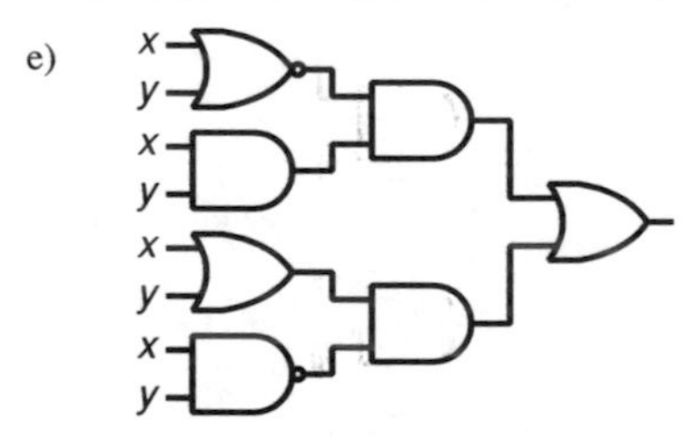

7.

a)

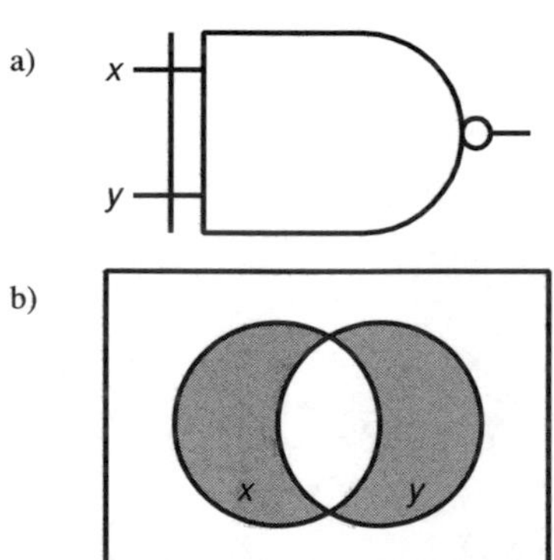

b)

c)

x	y	$\overline{(x \odot y)}$
0	0	0
0	1	1
1	0	1
1	1	0

d) $\overline{[(\bar{x} \wedge \bar{y}) \vee (x \wedge y)]}$ or because $\overline{(x \odot y)} = (x \oplus y)$, $\bar{x} \wedge y \vee (x \wedge \bar{y})$

e) 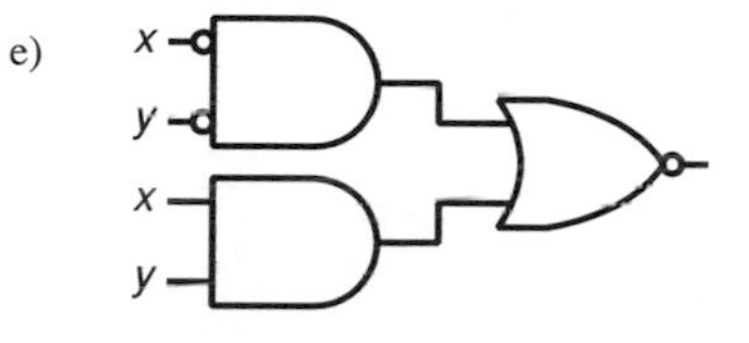

8.

a)

x	y	z	$(x \oplus y) \downarrow z$
0	0	0	1
0	0	1	0
0	1	0	0
0	1	1	0
1	0	0	0
1	0	1	0
1	1	0	1
1	1	1	0

b) $(x \oplus y) \downarrow z = \overline{[(x \oplus y) \vee z]} = \overline{[((\bar{x} \wedge y) \vee (x \wedge y)) \vee z]} =$
$[((((x \uparrow x) \uparrow y) \uparrow (x \uparrow (y \uparrow y))) \uparrow (((x \uparrow x) \uparrow y) \uparrow (x \uparrow (y \uparrow y)))] \uparrow$
$(z \uparrow z)) \uparrow [((((x \uparrow x) \uparrow y) \uparrow (x \uparrow (y \uparrow y))) \uparrow (((x \uparrow x) \uparrow y) \uparrow (x \uparrow (y \uparrow y)))) \uparrow (z \uparrow z)]$

9.

a)

x	y	$\bar{x} \vee \bar{y}$
0	0	1
0	1	1
1	0	1
1	1	0

b) $\bar{x} \vee \bar{y} = (\bar{\bar{x}}) \uparrow (\bar{\bar{y}}) = x \uparrow y$

10.

x	y	$x \rightarrow \bar{y}$
0	0	1
0	1	1
1	0	1
1	1	0

b) $x \rightarrow \bar{y} = \bar{x} \vee \bar{y} = \bar{\bar{x}} \uparrow \bar{\bar{y}} = x \uparrow y$

11.

a)

x	y	$(x \rightarrow y) \wedge (x \vee y)$
0	0	0
0	1	1
1	0	0
1	1	1

b) $(x \rightarrow y) \wedge (x \vee y) = [\overline{(x \uparrow \bar{y}) \uparrow (\bar{x} \uparrow \bar{y})}] =$
$\overline{[(x \uparrow (y \uparrow y)) \uparrow ((x \uparrow x) \uparrow (y \uparrow y))]} =$
$((x \uparrow (y \uparrow y)) \uparrow ((x \uparrow x) \uparrow (y \uparrow y))) \uparrow ((x \uparrow (y \uparrow y)) \uparrow ((x \uparrow x) \uparrow (y \uparrow y)))$ (or alternatively y)

12.

$\bar{x} = x \downarrow x$
$x \vee y = \overline{(x \downarrow y)} = (x \downarrow y) \downarrow (x \downarrow y)$
$x \wedge y = \bar{x} \downarrow \bar{y} = (x \downarrow x) \downarrow (y \downarrow y)$

Chapter 5

Exercise 5.1

1.

a) $(\bar{x}\bar{y}) \vee (\bar{x}y) \vee (xy)$

b)

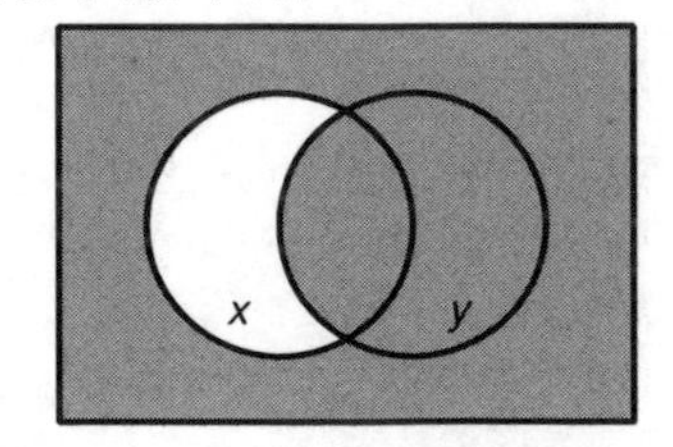

c)

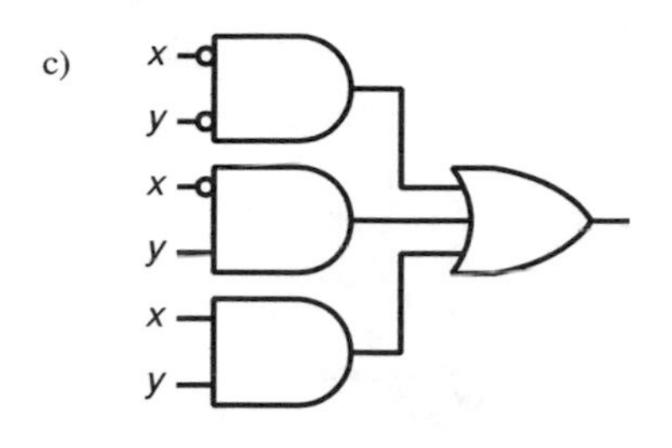

2.

a) $(\bar{a}\bar{b}\bar{c}d) \vee (\bar{a}\bar{b}c\bar{d}) \vee (\bar{a}b\bar{c}\bar{d}) \vee (\bar{a}b\bar{c}d) \vee (a\bar{b}\bar{c}d) \vee (a\bar{b}cd) \vee (ab\bar{c}\bar{d}) \vee (ab\bar{c}d)$

b)

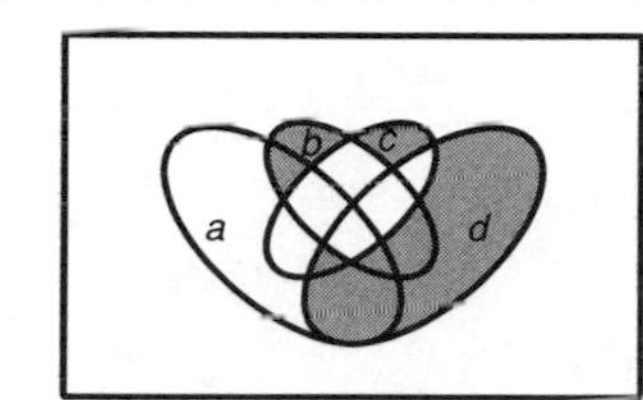

c)

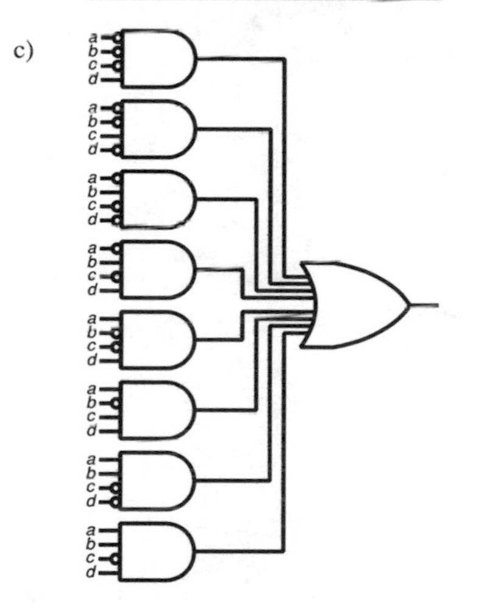

3.

a) $(\bar{x}\bar{y}z) \vee (\bar{x}y\bar{z}) \vee (x\bar{y}\bar{z})$

b)

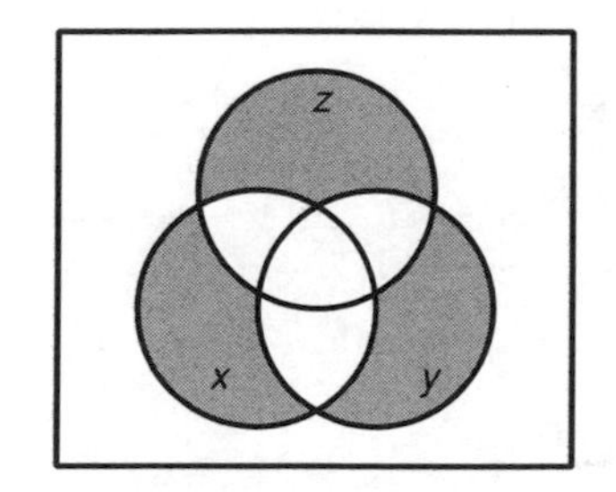

c)

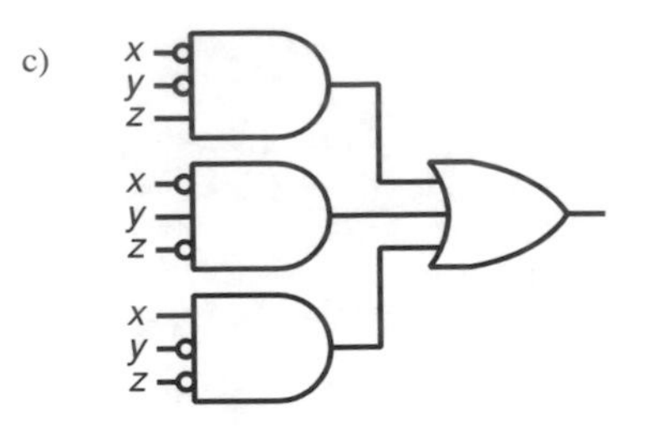

4.

a) $(\bar{x}\bar{y}\bar{z}) \vee (\bar{x}\bar{y}z) \vee (\bar{x}y\bar{z}) \vee (\bar{x}yz) \vee (x\bar{y}\bar{z}) \vee (x\bar{y}z) \vee (xyz)$

b)

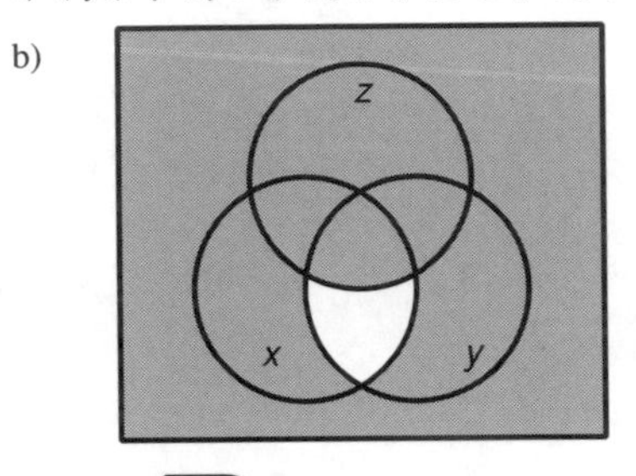

c)

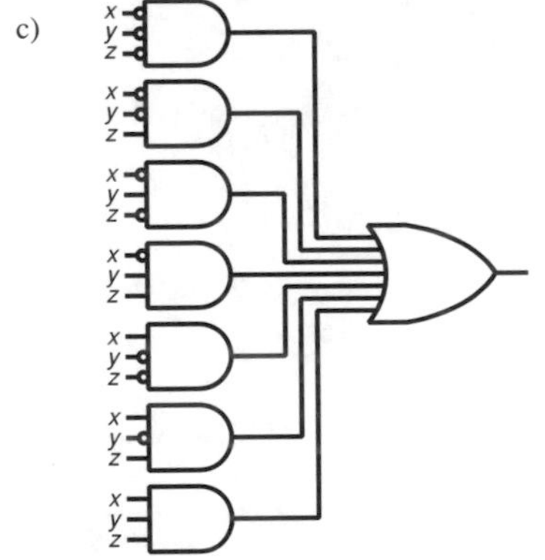

5. f0 (top bar) is on when we want the display to show 0, 2, 3, 5, 7, 8, and 9. So f0 $= (\bar{a}\bar{b}\bar{c}\bar{d}) \vee (\bar{a}\bar{b}c\bar{d}) \vee (\bar{a}\bar{b}cd) \vee (\bar{a}b\bar{c}d) \vee (\bar{a}bcd) \vee (a\bar{b}\bar{c}\bar{d}) \vee (a\bar{b}\bar{c}d)$

f1 (middle bar) is on when we want the display to show 2, 3, 4, 5, 6, 8, and 9. So f1 $= (\bar{a}\bar{b}c\bar{d}) \vee (\bar{a}\bar{b}cd) \vee (\bar{a}b\bar{c}\bar{d}) \vee (\bar{a}b\bar{c}d) \vee (\bar{a}bc\bar{d}) \vee (a\bar{b}\bar{c}\bar{d}) \vee (a\bar{b}\bar{c}d)$

f2 (bottom bar) is on when we want the display to show 0, 2, 3, 5, 6, and 8. So f2 $= (\bar{a}\bar{b}\bar{c}\bar{d}) \vee (\bar{a}\bar{b}c\bar{d}) \vee (\bar{a}\bar{b}cd) \vee (\bar{a}b\bar{c}d) \vee (\bar{a}bc\bar{d}) \vee (a\bar{b}\bar{c}\bar{d})$

f3 (upper right bar) is on when we want the display to show 0, 1, 2, 3, 4, 7, 8, and 9. So f3 $= (\bar{a}\bar{b}\bar{c}\bar{d}) \vee (\bar{a}\bar{b}\bar{c}d) \vee (\bar{a}\bar{b}c\bar{d}) \vee (\bar{a}\bar{b}cd) \vee (\bar{a}b\bar{c}\bar{d}) \vee (\bar{a}bcd) \vee (a\bar{b}\bar{c}\bar{d}) \vee (a\bar{b}\bar{c}d)$

f4 (upper left bar) is on when we want the display to show 0, 4, 5, 6, 8, and 9. So f4 $= (\bar{a}\bar{b}\bar{c}\bar{d}) \vee (\bar{a}b\bar{c}\bar{d}) \vee (\bar{a}b\bar{c}d) \vee (\bar{a}bc\bar{d}) \vee (a\bar{b}\bar{c}\bar{d}) \vee (a\bar{b}\bar{c}d)$

f5 (lower left bar) has already been done in the text

f6 (lower right bar) is on when we want the display to show 0, 1, 3, 4, 5, 6, 7, 8, and 9. So f6 $= (\bar{a}\bar{b}\bar{c}\bar{d}) \vee (\bar{a}\bar{b}\bar{c}d) \vee (\bar{a}\bar{b}cd) \vee (\bar{a}b\bar{c}\bar{d}) \vee (\bar{a}b\bar{c}d) \vee (\bar{a}bc\bar{d}) \vee (\bar{a}bcd) \vee (a\bar{b}\bar{c}\bar{d}) \vee (a\bar{b}\bar{c}d)$

Exercise 5.2

1.

a) $(\bar{x}\bar{y}) \vee (\bar{x}y) \vee (x\bar{y})$

b) $\bar{x} \vee \bar{y}$

2.

a) $\bar{x}\bar{y}$

b) $(x \vee \bar{y}) \wedge (\bar{x} \vee y) \wedge (\bar{x} \vee y)$

3.

a) $(\bar{x}\bar{y}) \vee (\bar{x}y) \vee (xy)$

b) $\bar{x} \vee y$

4.

a) xy

b) $(x \vee y) \wedge (x \vee \bar{y}) \wedge (\bar{x} \vee y)$

5.

a) $\bar{x} \vee y$

b)

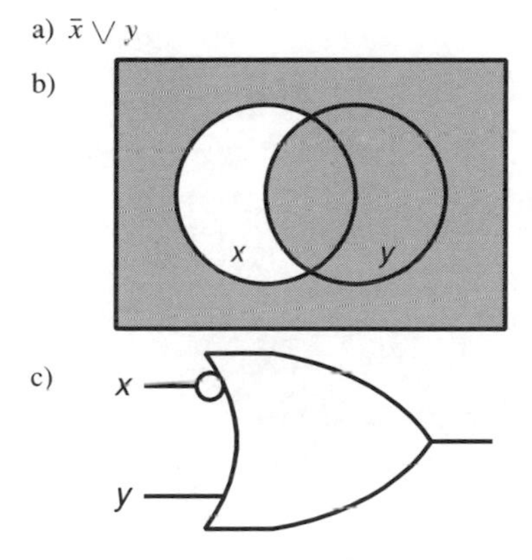

c)

6.

a) $(a \vee b \vee c \vee d) \wedge (a \vee b \vee \bar{c} \vee \bar{d}) \wedge (a \vee \bar{b} \vee \bar{c} \vee d) \wedge (a \vee \bar{b} \vee \bar{c} \vee \bar{d}) \wedge (\bar{a} \vee b \vee c \vee d) \wedge (\bar{a} \vee b \vee \bar{c} \vee d) \wedge (\bar{a} \vee \bar{b} \vee \bar{c} \vee d) \wedge (\bar{a} \vee \bar{b} \vee \bar{c} \vee \bar{d})$

b)

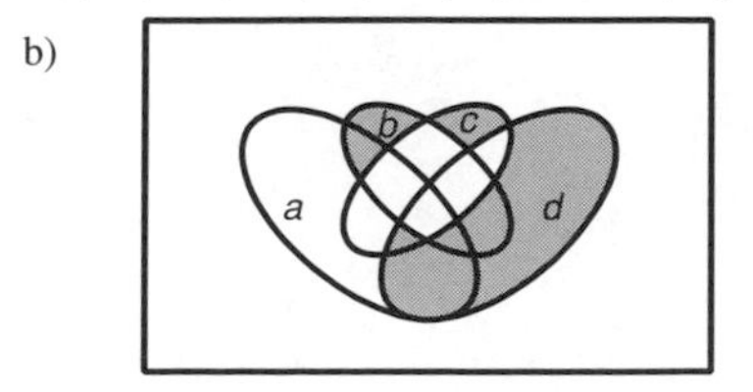

c) 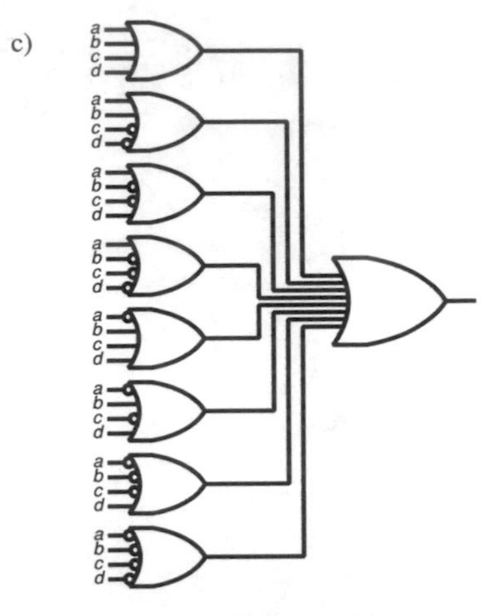

7.

a) $(x \vee y \vee z) \wedge (x \vee \bar{y} \vee \bar{z}) \wedge (\bar{x} \vee y \vee \bar{z}) \wedge (\bar{x} \vee \bar{y} \vee z) \wedge (\bar{x} \vee \bar{y} \vee \bar{z})$

b) 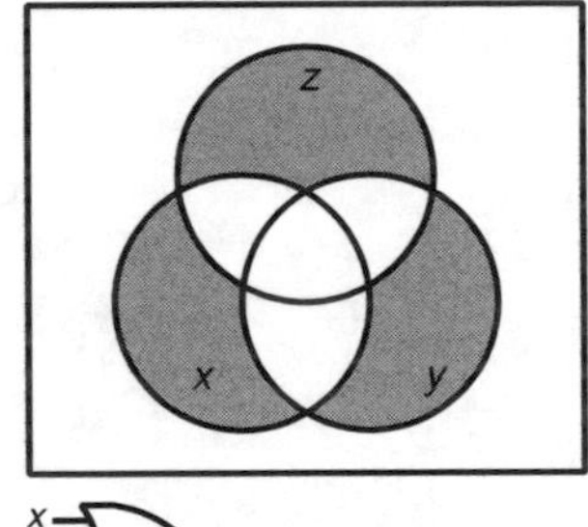

c) 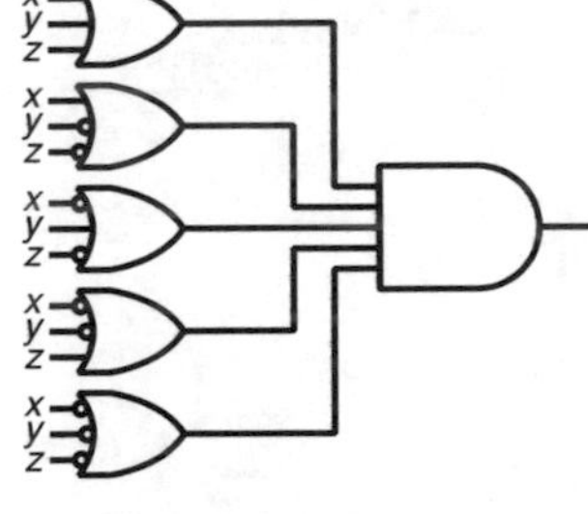

8.

a) $\bar{x} \vee \bar{y} \vee z$

b) 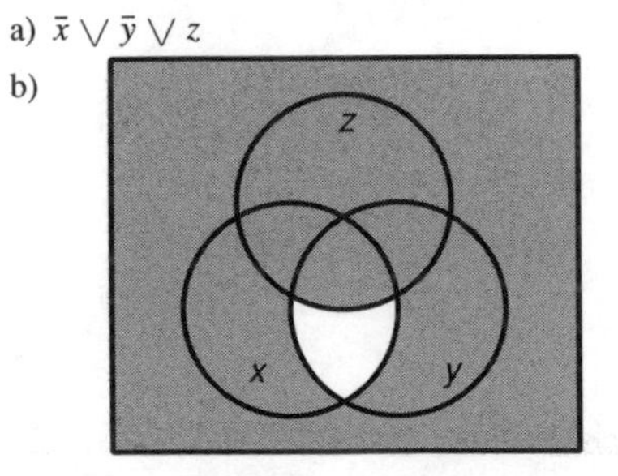

c) 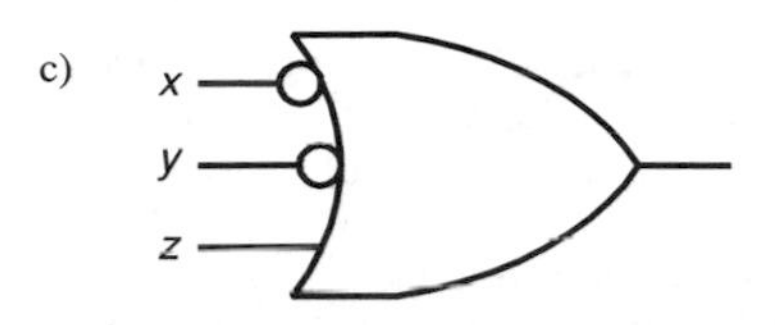

9. $(\bar{x}yz) \vee (x\bar{y}z) \vee (xy\bar{z}) \vee (xyz)$

10. f0 (top bar) is 0 when we want the display to show 1, 4, and 6. So f0 $= (a \vee b \vee c \vee \bar{d}) \wedge (a \vee \bar{b} \vee c \vee d) \wedge (a \vee \bar{b} \vee \bar{c} \vee d)$

    f1 (middle bar) is 0 when we want the display to show 0, 1, and 7. So f1 $= (a \vee b \vee c \vee d) \wedge (a \vee b \vee c \vee \bar{d}) \wedge (a \vee \bar{b} \vee \bar{c} \vee \bar{d})$

    f2 (bottom bar) is 0 when we want the display to show 1, 4, 7, and 9. So f2 $= (a \vee b \vee c \vee \bar{d}) \wedge (a \vee \bar{b} \vee c \vee d) \wedge (a \vee \bar{b} \vee \bar{c} \vee \bar{d}) \wedge (\bar{a} \vee b \vee c \vee \bar{d})$

    f3 (upper right bar) is 0 when we want the display to show 5 and 6. So f3 $= (a \vee \bar{b} \vee c \vee \bar{d}) \wedge (a \vee \bar{b} \vee \bar{c} \vee d)$

    f4 (upper left bar) is 0 when we want the display to show 1, 2, 3, and 7. So f4 $= (a \vee b \vee c \vee \bar{d}) \wedge (a \vee b \vee \bar{c} \vee d) \wedge (a \vee b \vee \bar{c} \vee \bar{d}) \wedge (a \vee \bar{b} \vee \bar{c} \vee \bar{d})$

    f5 (lower left bar) is 0 when we want the display to show 1, 3, 4, 5, 7, and 9. So f5 $= (a \vee b \vee c \vee \bar{d}) \wedge (a \vee b \vee \bar{c} \vee d) \wedge (a \vee \bar{b} \vee c \vee d) \wedge (a \vee \bar{b} \vee c \vee \bar{d}) \wedge (a \vee \bar{b} \vee \bar{c} \vee \bar{d}) \wedge (\bar{a} \vee b \vee c \vee \bar{d})$

    f6 (lower right bar) is 0 when we want the display to show a 2. So f6 $= (a \vee b \vee \bar{c} \vee d)$

11. Both of the functions m and t take 4 bits as input, so their truth tables are 16 lines long. Their truth tables are shown here:

a	b	c	d	m	t
0	0	0	0	0	0
0	0	0	1	0	0
0	0	1	0	0	0
0	0	1	1	d	1
0	1	0	0	0	0
0	1	0	1	d	1
0	1	1	0	d	1
0	1	1	1	1	0
1	0	0	0	0	0
1	0	0	1	d	1
1	0	1	0	d	1
1	0	1	1	1	0
1	1	0	0	d	1
1	1	0	1	1	0
1	1	1	0	1	0
1	1	1	1	1	0

If there are more than eight 1s in the output column of a function, it makes more sense to realize it using maxterms; if there are more than eight 0s, it makes more sense to realize it using minterms. Note that the tie function, t, is only 1 when there are exactly two 0s and two 1s among the 4-input bits. This is the case for the combinations $(a, b, c, d) = (0, 0, 1, 1); (0, 1, 0, 1); (0, 1, 1, 0); (1, 0, 0, 1); (1, 0, 1, 0); (1, 1, 0, 0)$. Because this is only six combinations, it makes sense to realize the tie function as a minterm function. $t = (\bar{a}\bar{b}cd) \vee (\bar{a}b\bar{c}d) \vee (\bar{a}bc\bar{d}) \vee (a\bar{b}\bar{c}d) \vee (a\bar{b}c\bar{d}) \vee (ab\bar{c}\bar{d})$. According to the specification of the function, m is a don't care in these six cases. The remaining 10 cases are split evenly: m is 1 in 5 cases and 0 in the other 5. We could realize m as a minterm function with five terms and let the don't care cases be 0, or we could realize m as a maxterm function and let the don't care cases be 1. Let us realize it as a minterm function. It is to be 1 when there are more 1s than 0s among the 4-input bits, that is whenever there are three 1s and only one 0, or when all 4-input bits are 1. So $m = (\bar{a}bcd) \vee (a\bar{b}cd) \vee (ab\bar{c}d) \vee (abc\bar{d}) \vee (abcd)$.

Chapter 6

Exercise 6.1

1.

a)

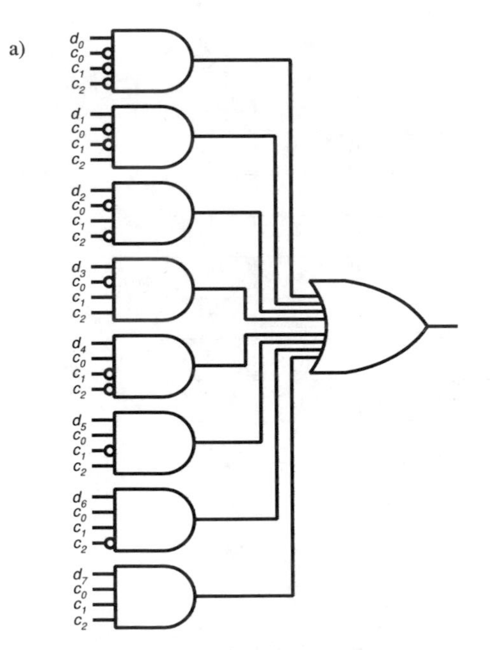

b) $(d_0\bar{c}_0\bar{c}_1\bar{c}_2) \vee (d_1\bar{c}_0\bar{c}_1c_2) \vee (d_2\bar{c}_0c_1\bar{c}_2) \vee (d_3\bar{c}_0c_1c_2) \vee (d_4c_0\bar{c}_1\bar{c}_2) \vee (d_5c_0\bar{c}_1c_2) \vee (d_6c_0c_1\bar{c}_2) \vee (d_7c_0c_1c_2)$

2. The truth table for a 1-control, 2-data multiplexer is as follows:

d_0	d_1	c	
0	0	0	0
0	0	1	0
0	1	0	0
0	1	1	1
1	0	0	1
1	0	1	0
1	1	0	1
1	1	1	1

So the minterm realization is $(\bar{d}_0 d_1 c \vee (d_0 \bar{d}_1 \bar{c}) \vee (d_0 d_1 \bar{c}) \vee (d_0 d_1 c)$ and the maxterm realization is $(d_0 \vee d_1 \vee c) \wedge (d_0 \vee d_1 \vee c) \wedge (d_0 \vee \bar{d}_1 \vee c) \wedge (d_0 \vee d_1 \vee \bar{c})$.

3. OR: Like the AND, it functions as a controllable gate, except that if the control input is 1, it blocks the data, producing a constant 1. Only if the control input is 0 will the output mirror the data input.

   NAND: If the control input is 0, the NAND blocks the data input and produces constant 1 as output. If the control input is 1, the output is the inverse (NOT) of the data input.

   NOR: If the control input is 1, NOR blocks the data input and just produces constant 0. If the control input is 0, the output is the inverse (NOT) of the data input.

   XOR: functions as a controllable NOT gate in that if the control input is 0, the data input is passed through to the output unchanged, but if the control input is 1, the output is the inverse (NOT) of the data input.

   COIN: also functions as a controllable NOT gate but with the logic of the control input inverted. If the control input is 1, the data input is passed through to the output unchanged, but if the control input is 0, the output is the inverse (NOT) of the data input.

Exercise 6.2

1. Referring to the truth table for the single-bit sum and carry functions, we see that there are 3 input bits (a, b, and c_i) and thus eight rows. Of these eight rows there are four 0s and four 1s in the output columns of both the sum and carry functions, so the minterm and maxterm realizations of each are equally optimal.

   a) c_o (carry out) $= (\bar{a}bc_i) \vee (a\bar{b}c_i) \vee (ab\bar{c}_i) \vee (abc_i)$

   s (sum) $= (\bar{a}\bar{b}c_i) \vee (\bar{a}b\bar{c}_i) \vee (a\bar{b}\bar{c}_i) \vee (abc_i)$

   b) c_o (carry out) $= (a \vee b \vee c_i) \wedge (a \vee b \vee \bar{c}_i) \wedge (a \vee \bar{b} \vee c_i) \wedge (\bar{a} \vee b \vee c_i)$

   s (sum) $= (a \vee b \vee c_i) \wedge (a \vee \bar{b} \vee \bar{c}_i) \wedge (\bar{a} \vee b \vee \bar{c}_i) \wedge (\bar{a} \vee \bar{b} \vee c_i)$

2. As before, we will design one cell of a cascaded comparator and chain an arbitrary number of these together. Also as before, the single cell will take input bits a and b (individual bits from the two multibit numbers being compared), e_i (A and B are equal so far), and ag_i (A is greater than B so far). Each cell will produce two output bits to be fed into the next cell, e_o and ag_o. Because we are now going upstream (i.e., from less significant bit positions to more significant ones), we cannot decide early on that A is greater or less than B and just pass the result down

the line. A can be greater than B up to the last cell, but because that last cell is the most significant bit if a is 0 and b is 1, then $A < B$.

We know that we will pass $(e, ag_i) = (0, 0)$ into the first cell; before we start, A and B are considered equal so far.

For a given cell, e_o is only to be true if e_i was true and $a = b$. So $e_o = e_i \wedge (a \odot b) = e_i \wedge [(\bar{a}\bar{b}) \vee (ab)]$.

Because we are cascading upstream, if a and b are different, that determines right there which is bigger up to this point, regardless of the information passed in by e_i and ag_i. If $a = b$, however, ag_o is true if and only if ag_i was true. So $ag_o = (a\bar{b}) \vee [(a \odot b) \wedge ag_i]$. That is, ag_o is true if a is 1 and b is 0 or if $a = b$ if ag_i is 1. Put in terms of AND, OR, and NOT, $ag_o = (a\bar{b}) \vee [((\bar{a}\bar{b}) \vee (ab)) \wedge ag_i]$.

The outputs of the last cell (the one comparing the bits in A and B that occupy the most significant bit positions) are taken as the output bits of the entire comparator.

Exercise 6.3

1. An ALU with two 6-bit input vectors and a 3-bit control vector has 6 + 6 + 3 or 15 individual input bits. Its truth table then would be 2^{15} or 32,768 lines long. Because the output vector is 6 bits wide as well, we would need six individual functions of 15-input bits each to realize this ALU.

2. Let us imagine an 8-to-3 encoder, and call the 8 input bits d_0-d_7 and the 3 output bits a, b, and c. The output bits are to be read in the order $[a, b, c]$ so that a is the most significant bit. If, for example, d_3 is 1 (and all other inputs are 0, as per our original assumption), then we want $(a, b, c) = (0, 1, 1)$ because $3_{10} = 011_2$. It should be clear that we want c to be 1 for every other input, specifically the odd ones. So $c = (d_1 \vee d_3 \vee d_5 \vee d_7)$. Similarly, we want b to be 1 for each input whose number when divided by 2 yields an odd number, i.e., 2, 3, 6, and 7. So $b = (d_2 \vee d_3 \vee d_6 \vee d_7)$. We want c to be 1 for each input whose number when divided by 4 yields an odd number, i.e., 4, 5, 6, and 7. So $c = (d_4 \vee d_5 \vee d_6 \vee d_7)$. This reasoning can be extended to any number of input bits.

3. Each output line is to be the AND of the data input and that particular combination of control inputs which selects that output line. So, for example, output line 5 is to be whatever the data input bit is ANDed with the combination $(c_0, c_1, c_2) = (1, 0, 1)$. So output $5 = d \wedge (c_0\bar{c}_1c_2)$. This reasoning is easily extended to the other output lines or for a demultiplexer of any number of output bits (as long as the control vector is correspondingly wide).

4. The trick is that a division by 2 is simply a shift one position to the right: $1001101 \div 2 = 0100110$. A division by 4 is a shift two positions to the right: $1001101 \div 4 = 0010011$. For our purposes, a shift function (and hence, a divider by 2) is a circuit that simply connects the input lines to the output lines but shifted over one position, with a 0 put out on the most significant output line and the least significant input line ignored:

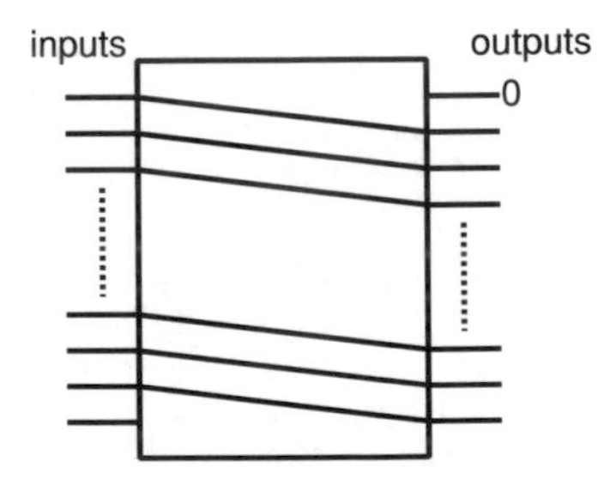

A circuit to shift by two positions just connects the input lines to the output lines but shifted over by two positions. This is the divider by four. The ALU then looks like this (all vectors are 8 bits wide except the 2-bit control vector):

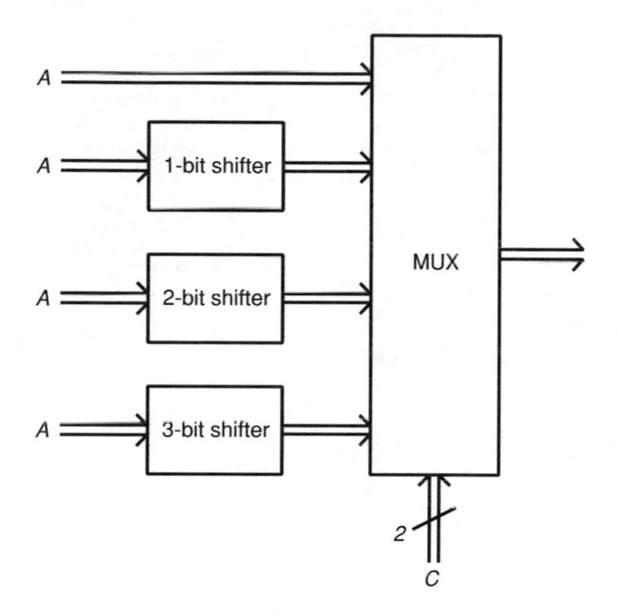

Chapter 7

Exercise 7.1

1. $1 \wedge x = x$:

x	$1 \wedge x$
0	0
1	1

$0 \wedge x = 0$:

x	$0 \wedge x$
0	0
1	0

2. $x \vee (y \wedge z) = (x \vee y) \wedge (x \vee z)$

x	y	z	$y \wedge z$	$x \vee (y \wedge z)$	$x \vee y$	$x \vee z$	$(x \vee y) \wedge (x \vee z)$
0	0	0	0	0	0	0	0
0	0	1	0	0	0	1	0
0	1	0	0	0	1	0	0
0	1	1	1	1	1	1	1
1	0	0	0	1	1	1	1
1	0	1	0	1	1	1	1
1	1	0	0	1	1	1	1
1	1	1	1	1	1	1	1

3. The Commutative Law of AND
4. The Special Property of 0 and AND
5. The Distributive Law of OR over AND
6. The Law of Involution
7. Special Properties of 0 and 1:

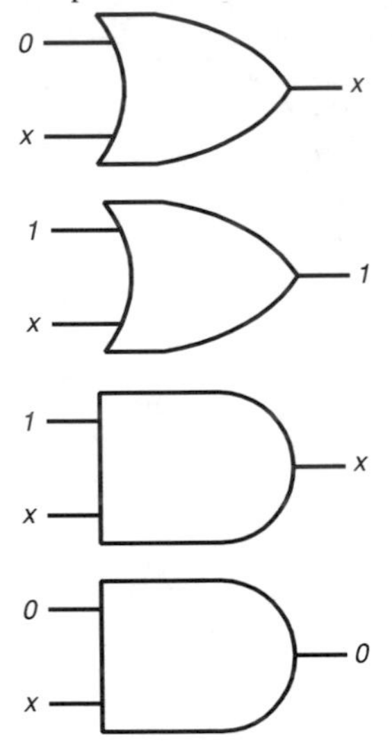

Complementation Laws:

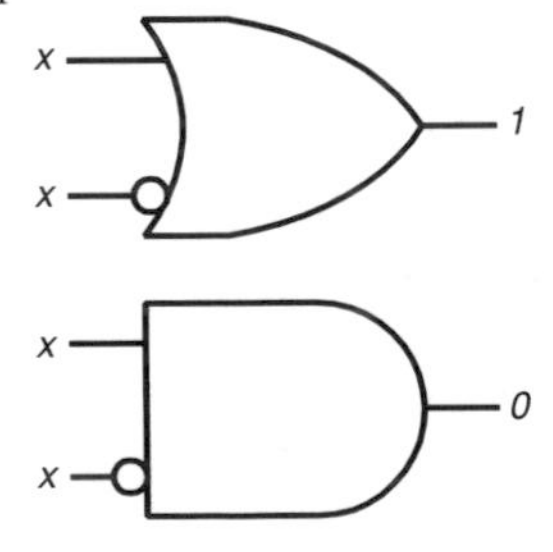

Law of Involution:

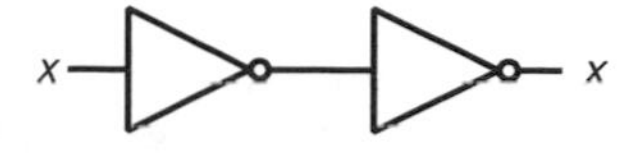

Commutative Laws:

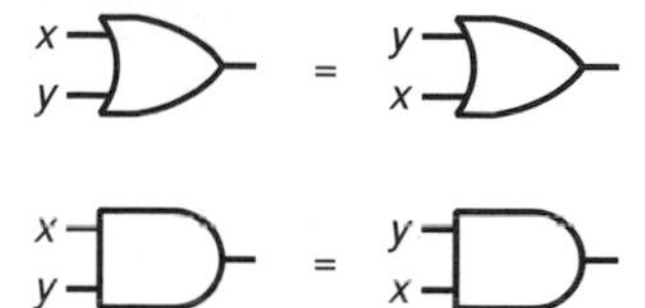

Distributive Law of AND over OR:

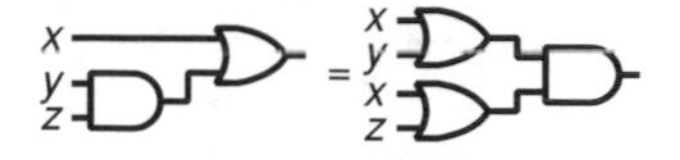

8.

a) x

b) $y \cap z$

c) $x \cup (y \cap z)$

d) $x \cup y$

e) $x \cup z$

f) $(x \cup y) \cap (x \cup z)$

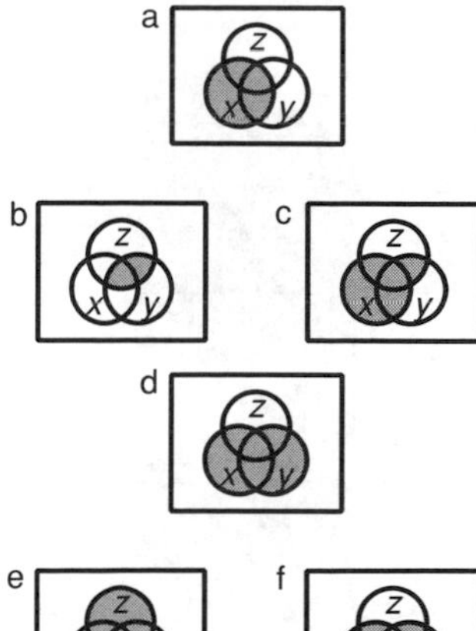

Exercise 7.2

1. $x \vee x = x$:

x	$x \vee x$
0	0
1	1

$x \wedge x = x$:

x	$x \wedge x$
0	0
1	1

2. $(x \vee y) \vee z = x \vee (y \vee z)$:

x	y	z	$x \vee y$	$(x \vee y) \vee z$	$y \vee z$	$x \vee (y \vee z)$
0	0	0	0	0	0	0
0	0	1	0	1	1	1
0	1	0	1	1	1	1
0	1	1	1	1	1	1
1	0	0	1	1	0	1
1	0	1	1	1	1	1
1	1	0	1	1	1	1
1	1	1	1	1	1	1

$(x \wedge y) \wedge z = x \wedge (y \wedge z)$:

x	y	z	$x \wedge y$	$(x \wedge y) \wedge z$	$y \wedge z$	$x \wedge (y \wedge z)$
0	0	0	0	0	0	0
0	0	1	0	0	0	0
0	1	0	0	0	0	0
0	1	1	0	0	1	0
1	0	0	0	0	0	0
1	0	1	0	0	0	0
1	1	0	1	0	0	0
1	1	1	1	1	1	1

3. $x \vee (x \wedge y) = x$:

x	y	$x \wedge y$	$x \vee (x \wedge y)$
0	0	0	0
0	1	0	0
1	0	0	1
1	1	1	1

$x \wedge (x \vee y) = x$:

x	y	$x \vee y$	$x \wedge (x \vee y)$
0	0	0	0
0	1	1	0
1	0	1	1
1	1	1	1

$x \vee (\bar{x} \wedge y) = x \vee y$:

x	y	$x \vee y$	$\bar{x}$	$\bar{x} \wedge y$	$x \vee (\bar{x} \wedge y)$
0	0	0	1	0	0
0	1	1	1	1	1
1	0	1	0	0	1
1	1	1	0	0	1

$x \wedge (\bar{x} \vee y) = x \wedge y$:

x	y	$x \wedge y$	$\bar{x}$	$\bar{x} \vee y$	$x \wedge (\bar{x} \vee y)$
0	0	0	1	1	0
0	1	0	1	1	0
1	0	0	0	0	0
1	1	1	0	1	1

$(x \wedge y) \vee (x \wedge \bar{y}) = x$:

x	y	$x \wedge y$	$\bar{y}$	$x \wedge \bar{y}$	$(x \wedge y) \vee (x \wedge \bar{y})$
0	0	0	1	0	0
0	1	0	0	0	0
1	0	0	1	1	1
1	1	1	0	0	1

$(x \vee y) \wedge (x \vee \bar{y}) = x$:

x	y	$x \vee y$	$\bar{y}$	$x \vee \bar{y}$	$(x \vee y) \wedge (x \vee \bar{y})$
0	0	0	1	1	0
0	1	1	0	0	0
1	0	1	1	1	1
1	1	1	0	1	1

4.

a) x

b) $x \cap y$

c) $x \cup (x \cap y)$

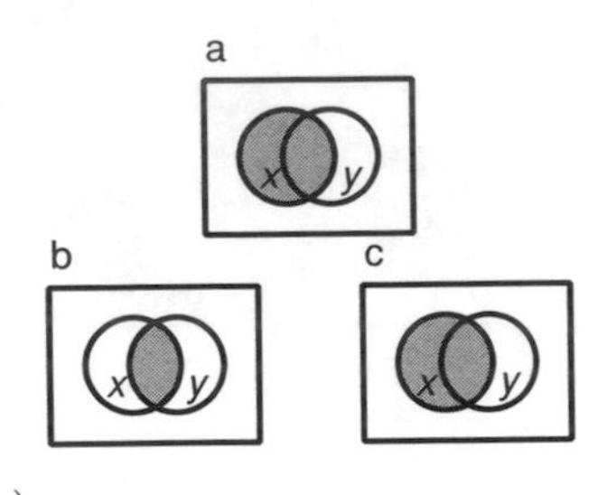

a) x

b) $x \cup y$

c) $x \cap (x \cup y)$

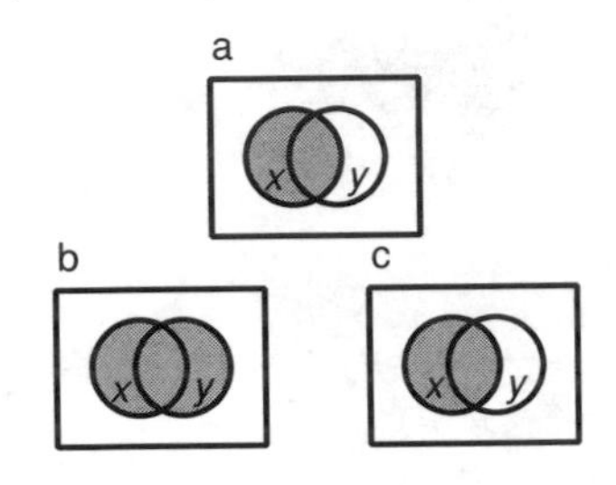

a) x

b) x'

c) $x' \cap y$

d) $x \cup (x' \cap y)$

e) $x \cup y$

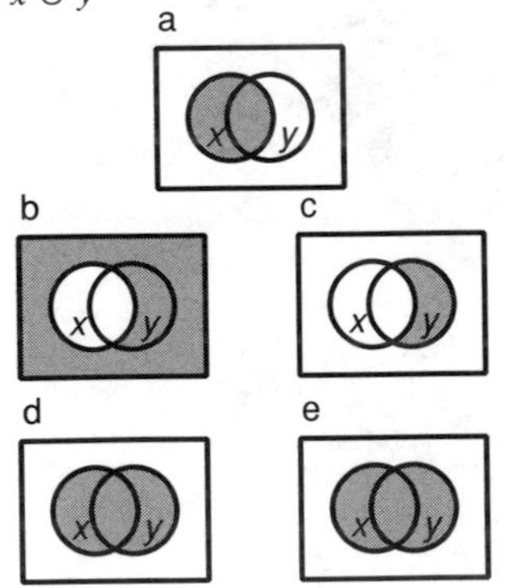

a) x

b) x'

c) $x' \cup y$

d) $x \cap (x' \cup y)$

e) $x \cap y$

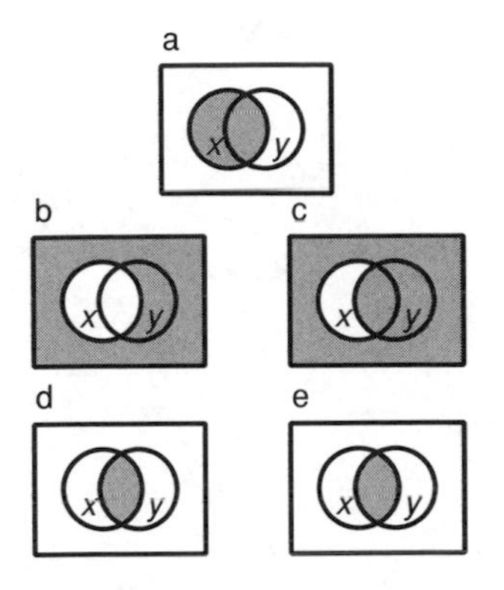

a) $x \cap y$

b) y'

c) $x \cap y'$

d) $(x \cap y) \cup (x \cap y')$

e) x

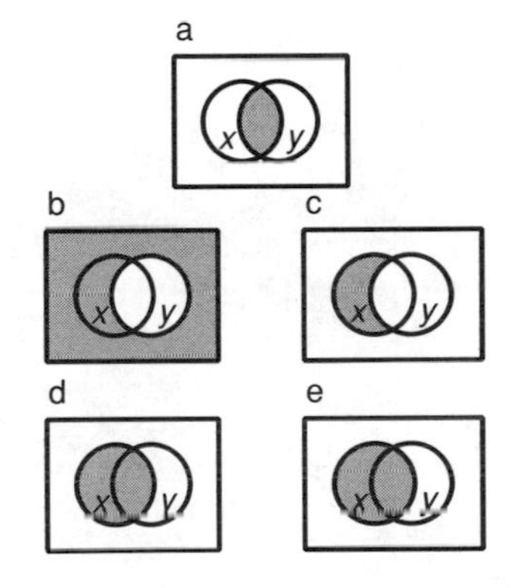

a) $x \cup y$

b) y'

c) $x \cup y'$

d) $(x \cup y) \cap (x \cup y')$

e) x

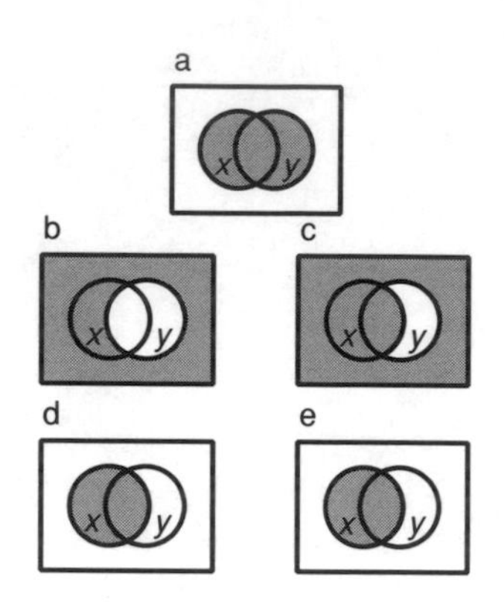

5.

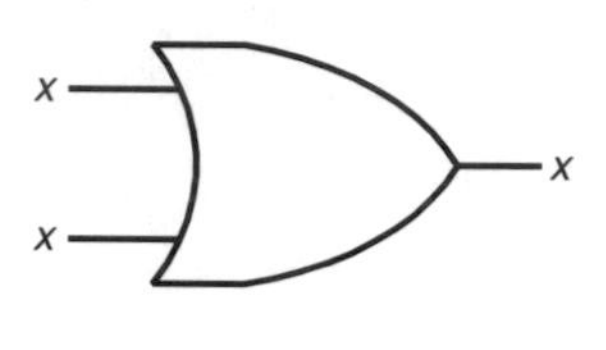

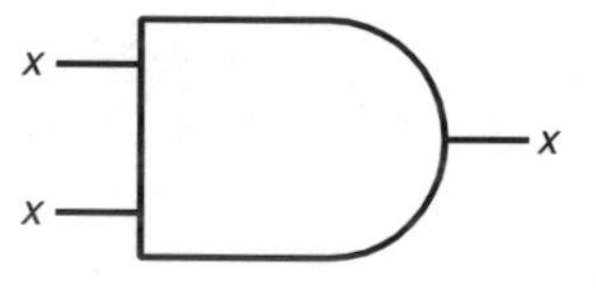

6.

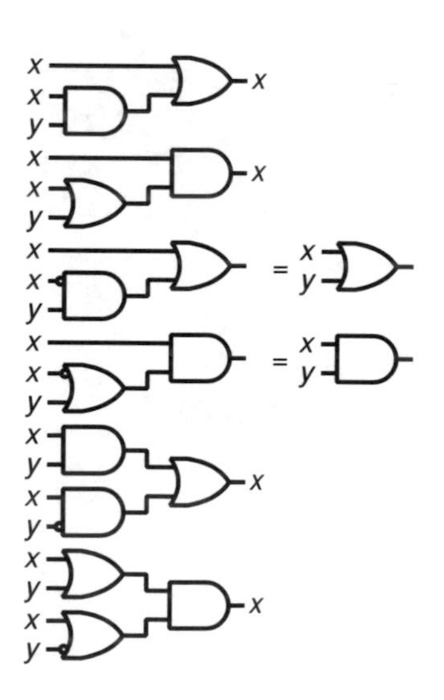

7.

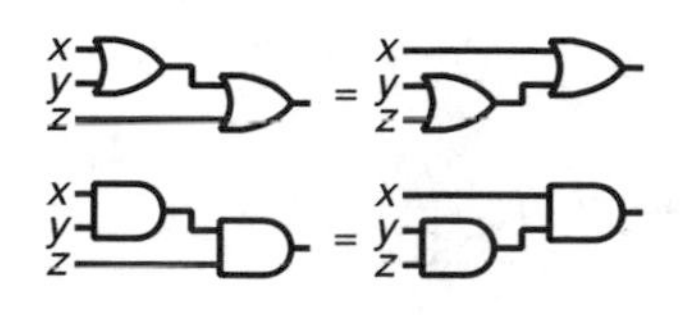

8. $x \wedge x =$
   $0 \vee (x \wedge x) =$ Special Property of 0 and OR
   $(x \wedge \bar{x}) \vee (x \wedge x) =$ Complementation Law of AND
   $x \vee (\bar{x} \wedge x) =$ Distributive Law of OR over AND
   $x \vee (x \wedge \bar{x}) =$ Commutativity Law of AND
   $x \vee 0 =$ Complementation Law of AND
   x Special Property of 0 and AND

9. $x \wedge (x \vee y) =$
   $(0 \vee x) \wedge (x \vee y) =$ Special Property of OR and 1
   $(x \vee 0) \wedge (x \vee y) =$ Commutativity Law of OR
   $x \vee (0 \wedge y) =$ Distributivity Law of OR over AND
   $x \vee 0 =$ Special Property of AND and 0
   x Special Property of OR and 0

   $x \vee (\bar{x} \wedge y) =$
   $(x \vee \bar{x}) \wedge (x \vee y) =$ Distributivity Law of OR over AND
   $1 \wedge (x \vee y) =$ Complementation Law of OR
   $x \vee y$ Special Property of AND and 1

   $x \wedge (\bar{x} \vee y) =$
   $(x \wedge \bar{x}) \vee (x \wedge y) =$ Distributivity Law of AND over OR
   $0 \vee (x \wedge y) =$ Complementation Law of AND
   $x \wedge y$ Special Property of OR and 0

   $(x \wedge y) \vee (x \wedge \bar{y}) =$
   $x \wedge (y \vee \bar{y}) =$ Distributivity Law of AND over OR
   $x \wedge 1 =$ Complementation Law of OR
   x Special Property of AND and 1

   $(x \vee y) \wedge (x \vee \bar{y}) =$
   $x \vee (y \wedge \bar{y}) =$ Distributivity Law of OR over AND
   $x \vee 0 =$ Complementation Law of AND
   x Special Property of OR and 0

10. $(x \vee y) \vee z =$
    $[(x \vee y) \vee z] \wedge 1 =$ Special Property of AND and 1
    $[(x \vee y) \vee z] \wedge (x \vee \bar{x}) =$ Complementation Law of OR
    $[((x \vee y) \vee z) \wedge x] \vee [((x \vee y) \vee z) \wedge \bar{x}] =$ Distributivity Law of AND over OR
    $[x \wedge ((x \vee y) \vee z)] \vee [\bar{x} \wedge ((x \vee y) \vee z)] =$ Commutativity Law of AND
    $[(x \wedge (x \vee y)) \vee (x \wedge z)] \vee [(\bar{x} \wedge (x \vee y)) \vee (\bar{x} \wedge z)] =$ Distributivity Law of AND over OR
    $[x \vee (x \wedge z)] \vee [(\bar{x} \wedge (x \vee y)) \vee (\bar{x} \wedge z)] =$ Absorption Law
    $[x \vee (x \wedge z)] \vee [((\bar{x} \wedge x) \vee (\bar{x} \wedge y)) \vee (\bar{x} \wedge z)] =$ Distributivity Law of AND over OR
    $[x \vee (x \wedge z)] \vee [(0 \vee (\bar{x} \wedge y)) \vee (\bar{x} \wedge z)] =$ Complementation Law of AND
    $[x \vee (x \wedge z)] \vee [(\bar{x} \wedge y) \vee (\bar{x} \wedge z)] =$ Special Property of OR and 0

$x \vee [(\bar{x} \wedge y) \vee (\bar{x} \wedge z)]$ = Absorption Law
$x \vee [\bar{x} \wedge (y \vee z)]$ = Distributivity Law of AND over OR
$x \vee (y \vee z)$ Absorption Law

11. $(x \vee y) \wedge (x \vee \bar{y}) \wedge (\bar{x} \vee y)$ =
$[(x \vee y) \wedge (x \vee \bar{y})] \wedge (\bar{x} \vee y)$ = Associative Law of AND
$[x \vee (y \wedge \bar{y})] \wedge (\bar{x} \vee y)$ = Distributive Law of OR over AND
$(x \vee 0) \wedge (\bar{x} \vee y)$ = Complementation Law of AND
$x \wedge (\bar{x} \vee y)$ = Special Property of OR and 0
$(x \wedge \bar{x}) \vee (x \wedge y)$ = Distributive Law of AND over OR
$0 \vee (x \wedge y)$ = Complementation Law of AND
$x \wedge y$ Special Property of OR and 0

12. $(\bar{x} \wedge \bar{y}) \vee (\bar{x} \wedge y) \vee (x \wedge y)$ =
$[(\bar{x} \wedge \bar{y}) \vee (\bar{x} \wedge y)] \vee (x \wedge y)$ = Associativity Law of OR
$[\bar{x} \wedge (\bar{y} \vee y)] \vee (x \wedge y)$ = Distributive Law of AND over OR
$(\bar{x} \wedge 1) \vee (x \wedge y)$ = Complementation Law of OR
$\bar{x} \vee (x \wedge y)$ = Special Property of AND and 1
$(\bar{x} \vee x) \wedge (\bar{x} \vee y)$ = Distributive Law of OR over AND
$1 \wedge (\bar{x} \vee y)$ = Complementation Law of OR
$\bar{x} \vee y$ Special Property of AND and 1

13. $(\bar{x} \wedge \bar{y}) \vee (x \wedge y)$ =
$[(\bar{x} \wedge \bar{y}) \vee x] \wedge [(\bar{x} \wedge \bar{y}) \vee y]$ = Distributive Law of OR over AND
$[(x \vee \bar{x}) \wedge (x \vee \bar{y})] \wedge [(\bar{x} \vee y) \wedge (\bar{y} \vee y)]$ = Distributive Law of OR over AND
$[1 \wedge (x \vee \bar{y})] \wedge [(\bar{x} \vee y) \wedge 1]$ = Complementation Law of OR
$(x \vee \bar{y}) \wedge (\bar{x} \vee y)$ = Special Property of AND and 1

14. $\overline{(x \wedge x)}$ =
$(\bar{x} \vee \bar{x})$ = De Morgan
$\bar{x}$ Complementation Law of OR

15. $\overline{[\overline{(x \wedge x)} \wedge \overline{(y \wedge y)}]}$ =
$[\overline{\overline{(x \wedge x)}}] \vee [\overline{\overline{(y \wedge y)}}]$ = De Morgan
$(x \wedge x) \vee (y \wedge y)$ = Law of Involution
$x \vee y$ Idempotence

16. $\overline{(\overline{(x \wedge y)} \wedge \overline{(x \wedge y)})}$ =
$[\overline{\overline{(x \wedge y)}}] \vee [\overline{\overline{(x \wedge y)}}]$ = De Morgan
$(x \wedge y) \vee (x \wedge y)$ = Law of Involution
$x \wedge y$ Idempotence

17. $\overline{(a \wedge b \wedge c \wedge d)}$ =
$\overline{\overline{[(\underline{a \wedge b}) \wedge (\underline{c \wedge d})]}}$ = Associative Law of AND
$(a \wedge b) \vee (c \wedge d)$ = De Morgan
$(\bar{a} \vee \bar{b}) \vee (\bar{c} \vee \bar{d})$ = De Morgan
$\bar{a} \vee \bar{b} \vee \bar{c} \vee \bar{d}$ Associative Law of OR

18. $a \wedge (b \vee c \vee d)$ =
$a \wedge ((b \vee c) \vee d)$ = Associative Law of OR
$(a \wedge (b \vee c)) \vee (a \wedge d)$ = Distributive Law of AND over OR
$[(a \wedge b) \vee (a \wedge c)] \vee (a \wedge d)$ = Distributive Law of AND over OR
$(a \wedge b) \vee (a \wedge c) \vee (a \wedge d)$ Associative Law of OR

19. $M_1 \vee M_3 \vee M_5 \vee M_7 = (\bar{x} \wedge \bar{y} \wedge z) \vee (\bar{x} \wedge y \wedge z) \vee (x \wedge \bar{y} \wedge z) \vee (x \wedge y \wedge z) =$
    $[(\bar{x} \wedge \bar{y}) \wedge z] \vee [(\bar{x} \wedge y) \wedge z] \vee [(x \wedge \bar{y}) \wedge z] \vee [(x \wedge y) \wedge z] =$ Associative Law of AND
    $[((\bar{x} \wedge \bar{y}) \wedge z) \vee ((\bar{x} \wedge y) \wedge z)] \vee [((x \wedge \bar{y}) \wedge z) \vee ((x \wedge y) \wedge z)] =$ Associative Law of OR
    $[z \wedge ((\bar{x} \wedge \bar{y}) \vee (\bar{x} \wedge y))] \vee [z \wedge ((x \wedge \bar{y}) \vee (x \wedge y))] =$ Distributive Law of AND over OR
    $(z \wedge \bar{x}) \vee (z \wedge x) =$ Absorption Law
    $z =$ Absorption Law

20. Trick question; this function is as simple as it gets in minterm form.

21. We want some number of individual terms ANDed together, where each of these terms contains every variable (complemented or not) ORed together.

    $z \wedge (\bar{x} \vee y) =$
    $[(z \vee x) \wedge (z \vee \bar{x})] \wedge (\bar{x} \vee y) =$ Absorption Law
    $[(((z \vee x) \vee y) \wedge ((z \vee x) \vee \bar{y})) \wedge (z \vee \bar{x})] \wedge (\bar{x} \vee y) =$ Absorption Law
    $[(((z \vee x) \vee y) \wedge ((z \vee x) \vee \bar{y})) \wedge (((z \vee \bar{x}) \vee y) \wedge ((z \vee \bar{x}) \vee \bar{y}))] \wedge (\bar{x} \vee y) =$ Absorption Law
    $[(((z \vee x) \vee y) \wedge ((z \vee x) \vee \bar{y})) \wedge (((z \vee \bar{x}) \vee y) \wedge ((z \vee \bar{x}) \vee \bar{y}))] \wedge [((\bar{x} \vee y) \vee z) \wedge ((\bar{x} \vee y) \vee \bar{z})] =$ Absorption Law
    $[((z \vee x \vee y) \wedge (z \vee x \vee \bar{y})) \wedge ((z \vee \bar{x} \vee y) \wedge (z \vee \bar{x} \vee \bar{y}))] \wedge [(\bar{x} \vee y \vee z) \wedge (\bar{x} \vee y \vee \bar{z})]$
    $=$ Associative Law of OR
    $(z \vee x \vee y) \wedge (z \vee x \vee \bar{y}) \wedge (z \vee \bar{x} \vee y) \wedge (z \vee \bar{x} \vee \bar{y}) \wedge (\bar{x} \vee y \vee z) \wedge (\bar{x} \vee y \vee \bar{z}) =$ Associative Law of AND
    $(x \vee y \vee z) \wedge (x \vee \bar{y} \vee z) \wedge (\bar{x} \vee y \vee z) \wedge (\bar{x} \vee \bar{y} \vee z) \wedge (\bar{x} \vee y \vee z) \wedge (\bar{x} \vee y \vee \bar{z}) =$ Commutative Law of OR (shown explicitly to get the function in the usual form in which the variables appear in alphabetical order)

22. Note that this is the implication function
    $\overline{(x \vee y)} \vee y =$
    $(\bar{x} \wedge \bar{y}) \vee y =$ De Morgan
    $\bar{x} \vee y =$ Absorption Law
    $\bar{x} \vee (x \wedge y) =$ Absorption Law
    $(\bar{x} \wedge y) \vee (\bar{x} \wedge \bar{y}) \vee (x \wedge y) =$ Absorption Law

23. $\overline{(x \vee y)} \vee y =$
    $(\bar{x} \wedge \bar{y}) \vee y =$ De Morgan
    $\bar{x} \vee y =$ Absorption Law

Exercise 7.3

1. Not constructed out of NOR: $\overline{(x \vee x)} = \bar{x}$
   AND constructed out of NOR: $\overline{[\overline{(x \vee x)} \vee \overline{(y \vee y)}]} = x \wedge y$
   OR constructed out of NOR: $\overline{[\overline{(x \vee y)} \vee \overline{(x \vee y)}]} = x \vee y$

Chapter 8

Exercise 8.1

1. No consensus exists; no opposition
2. $abcf$
3. $wy\bar{z}$
4. $\bar{b}cde\bar{f}$
5.
   a) $a \cap x$

   b) $b \cap x'$

   c) $a \cap b$

   d) $(a \cap x) \cup (b \cap x')$

   e) $(a \cap x) \cup (b \cap x') \cup (a \cap b)$

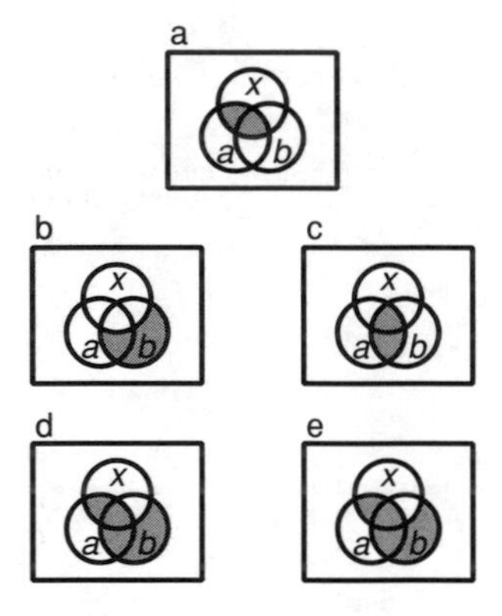

6. $(a \vee x) \wedge (b \vee \bar{x}) = (a \vee x) \wedge (b \vee \bar{x}) \wedge (a \vee b)$
7. $[bd \wedge (a \vee \bar{c})] \vee \bar{a} \vee a\bar{d} = (abd \vee b\bar{c}d) \vee \bar{a} \vee a\bar{d} =$ Distributive Law of AND over OR
   $abd \vee b\bar{c}d \vee \bar{a} \vee a\bar{d} =$ Associative Law of OR

Now the function is in SOP form. With all consensuses ORed in, it becomes $abd \vee b\bar{c}d \vee \bar{a} \vee a\bar{d} \vee bd \vee ab \vee ab\bar{c} \vee \bar{d}$.

a	b	c	d	$(bd \wedge (a \vee \bar{c})) \vee \bar{a} \vee a\bar{d}$	$abd \vee b\bar{c}d \vee \bar{a} \vee a\bar{d} \vee bd \vee ab \vee ab\bar{c} \vee \bar{d}$
0	0	0	0	1	1
0	0	0	1	1	1
0	0	1	0	1	1
0	0	1	1	1	1
0	1	0	0	1	1
0	1	0	1	1	1
0	1	1	0	1	1
0	1	1	1	1	1
1	0	0	0	1	1
1	0	0	1	0	0
1	0	1	0	1	1
1	0	1	1	0	0
1	1	0	0	1	1
1	1	0	1	1	1
1	1	1	0	1	1
1	1	1	1	1	1

Exercise 8.2

1. $\bar{x} \vee \bar{y}$
2. $bc \vee \bar{c}d \vee a\bar{c} \vee bd \vee ab \vee \bar{a}d$
3. $\bar{y}$
4. $ac \vee cd$
5. $\bar{a}\bar{b}c_i \vee \bar{a}b\bar{c}_i \vee a\bar{b}\bar{c}_i \vee abc_i$
6. $\bar{a}bc_i \vee a\bar{b}c_i \vee ab\bar{c}_i \vee abc_i$

Exercise 8.3

1. h: rides a hover scooter
   c: resident of Cathode City
   j: owns a jet pack

e: is eerily happy

$\bar{h} \rightarrow \bar{c}$
$\bar{j} \rightarrow h$
$je \rightarrow c$

$\bar{h}c \vee \bar{j}\bar{h} \vee je\bar{c} = 0$

which is, in BCF: $\bar{h}c \vee \bar{j}\bar{h} \vee je\bar{c} \vee e\bar{h} = 0$

The original premises are unchanged. The new prime implicant, $e\bar{h}$, may be taken as $e \rightarrow h$ or "Anyone who is eerily happy rides a hover scooter."

2. m: murples
g: gleebs
p: phaetinos

$m \rightarrow g$
$p \rightarrow \bar{g}$
$\bar{m} \rightarrow p$

$m\bar{g} \vee pg \vee \bar{m}\bar{p} = 0$

which is, in BCF: $m\bar{g} \vee pg \vee \bar{m}\bar{p} \vee mp \vee \bar{g}\bar{p} \vee g\bar{m} = 0$

The original premises are unchanged. The new prime implicants mp, $\bar{g}\bar{p}$, and $g\bar{m}$ may be interpreted as, respectively, "No murple is a phaetino (or alternatively no phaetino is a murple)," "All nongleebs are phaetinos (or alternatively all nonphaetinos are gleebs)," and "All gleebs are murples."

3. r: red carpet
o: orange wallpaper
s: striped sofa

$r \rightarrow \bar{o}$
$s \rightarrow o$
$\bar{r} \rightarrow \bar{s}$

$ro \vee s\bar{o} \vee \bar{r}s = 0$

which is, in BCF: $ro \vee s = 0$

So all the premises boil down to "If the carpet is red, then the wallpaper must not be orange (and vice versa)" and "The sofa must not be striped."

4. b: buzzer sounds
l: light flashes
w: warning displayed on console
r: bell rings

$bl \rightarrow w$
$l \rightarrow r$
$w \rightarrow \bar{r}$

$bl\bar{w} \vee l\bar{r} \vee wr = 0$

which is, in BCF: $l\bar{r} \vee wr \vee bl \vee lw = 0$

The term $bl\bar{w}$ has been subsumed by the more general term bl, which may be interpreted as "When the buzzer sounds, the light is not flashing (and vice versa)." In addition, there is the new prime implicant, lw, which may be interpreted as "When the light is flashing, the warning is not displayed on the console (and vice versa)." Essentially, the light flashing is mutually exclusive with the buzzer and the warning. The first of the original premises then provides some information that helps us derive the final result but refers to an impossible situation (i.e., that of the light flashing and the buzzer sounding at the same time, as well as that of the light flashing and the warning being displayed at the same time).

5. c: carnivore
   r: must be held in a reinforced cage
   w: winged
   v: must be well ventilated
   b: must be bathed daily

   $c \rightarrow r$
   $w \rightarrow v$
   $b \rightarrow \bar{w}$
   $\bar{c} \rightarrow w$

   $c\bar{r} \vee w\bar{v} \vee bw \vee \bar{c}\bar{w} = 0$

   which is, in BCF: $c\bar{r} \vee w\bar{v} \vee bw \vee \bar{c}\bar{w} \vee \bar{r}\bar{w} \vee \bar{c}\bar{v} \vee \bar{r}\bar{v} \vee b\bar{c} \vee b\bar{r} = 0$

   The original terms survive unchanged; the original premises are thus prime implicants. There are, however, some new prime implicants as well: $\bar{r}\bar{w}$, $\bar{c}\bar{v}$, $\bar{r}\bar{v}$, $b\bar{c}$, and $b\bar{r}$. These may be interpreted as, respectively, "Every animal is either in a reinforced cage, has wings, or both (i.e., all non-winged creatures are in reinforced cages and all creatures not in reinforced cages have wings);" "Every animal is either a carnivore or resides in a well-ventilated cage or both;" "Every animal is either in a well-ventilated cage or a reinforced cage or both;" "All animals that must be bathed daily are carnivores;" and "All animals that must be bathed daily live in reinforced cages."

Exercise 8.4

1.

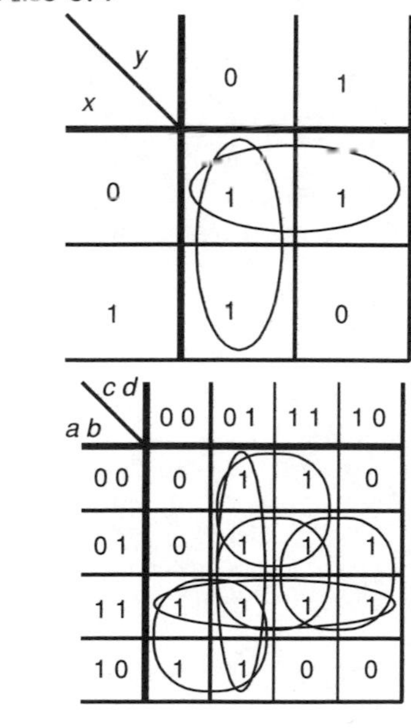

2.

3.

xy \ z	0	1
00	1	1
01	0	0
11	0	0
10	1	1

4.

ab \ cd	00	01	11	10
00	0	0	1	0
01	0	0	1	0
11	0	0	1	1
10	0	0	1	1

5.

ab \ cd	00	01	11	10
00	1	d	1	0
01	0	0	0	1
11	1	d	0	0
10	d	1	1	1

$ab\bar{c}\bar{d} \vee \bar{a}\bar{b}\bar{c}\bar{d} \vee \bar{a}bc\bar{d} \vee \bar{b}cd \vee a\bar{b}d \vee a\bar{b}c$

Index

A

B

C

D

E

G

H

I

J

K

L

M

N

O

P

Q

R

S

T

U

V

About the Author

John Gregg graduated from Trinity College in Hartford, Connecticut, with a B.S. in Computer Science. He was the first Computer Science graduate and the first Computer Science Fellow from Trinity. His initial exposure to Boolean algebra and digital circuits came in an undergraduate electrical engineering course. Struck by the simplicity and power of this system of logic, he, together with another undergraduate, taught the material to seventh and eighth grade students during a semester of independent study.

Since his graduation, he has worked for BBN (now owned by GTE) and Shiva Corporation as a computer programmer, specifically in the area of dedicated computer networking devices, such as packet switches and routers. In this capacity he programs in C and assembly language, implementing networking protocols, device drivers, and embedded operating systems code.

John Gregg, who also has occasionally worked as an actor and artist's model, lives in Cambridge, Massachusetts, with his wife, daughter, and Bert the cat.